# The Scott, Foresman PROCOM Series

**Series Editors**

**Roderick P. Hart**
University of Texas at Austin

**Ronald L. Applbaum**
Pan American University

## Titles in the PROCOM Series

BETTER WRITING FOR PROFESSIONALS
*A Concise Guide*
Carol Gelderman

BETWEEN YOU AND ME
*The Professional's Guide to Interpersonal Communication*
Robert Hopper
In consultation with Lillian Davis

COMMUNICATION STRATEGIES FOR TRIAL ATTORNEYS
K. Phillip Taylor
Raymond W. Buchanan
David U. Strawn

THE CORPORATE MANAGER'S GUIDE TO BETTER COMMUNICATION
W. Charles Redding
In consultation with Michael Z. Sincoff

THE ENGINEER'S GUIDE TO BETTER COMMUNICATION
Richard Arthur
In consultation with Volkmar Reichert

GETTING THE JOB DONE
*A Guide to Better Communication for Office Staff*
Bonnie Johnson
In consultation with Geri Sherman

THE GUIDE TO BETTER COMMUNICATION IN GOVERNMENT SERVICE
Raymond L. Falcione
In consultation with James G. Dalton

THE MILITARY OFFICER'S GUIDE TO BETTER COMMUNICATION
L. Brooks Hill
In consultation with Major Michael Gallagher

THE NURSE'S GUIDE TO BETTER COMMUNICATION
Robert E. Carlson
In consultation wih Margaret Kidwell Udin and Mary Carlson

THE PHYSICIAN'S GUIDE TO BETTER COMMUNICATION
Barbara F. Sharf
In consultation with Dr. Joseph A. Flaherty

THE POLICE OFFICER'S GUIDE TO BETTER COMMUNICATION
Keith V. Erickson
T. Richard Cheatham
In consultation with Frank Dyson

PROFESSIONALLY SPEAKING
*A Concise Guide*
Robert Doolittle
In consultation with Thomas Towers


For further information, write to

Professional Publishing Group
Scott, Foresman and Company
1900 East Lake Avenue
Glenview, IL 60025

# The Engineer's Guide to Better Communication

**SERIES EDITORS**

**Roderick P. Hart**
*University of Texas at Austin*

**Ronald L. Applbaum**
*Pan American University*

# The Engineer's Guide to Better Communication

**Dr. Richard Arthur**
University of Wisconsin at Milwaukee

in consultation with
Volkmar Reichert
City of Glendale, Wisconsin

**Scott, Foresman and Company    Glenview, Illinois**
Dallas, Texas   Oakland, New Jersey   Palo Alto, California
Tucker, Georgia   London

ACKNOWLEDGMENTS

From THE PROCESS OF COMMUNICATION by David K. Berlo. Copyright © 1960 by Holt, Rinehart and Winston, Inc. Reprinted by permission of Holt, Rinehart and Winston, CBS College Publishing.

From "Toward a Theory of Human Communication" by Frank E. X. Dance in HUMAN COMMUNICATION THEORY, edited by Frank E. X. Dance. Copyright © 1967 by Holt, Rinehart and Winston, Inc. Reprinted by permission of the author.

From the Introduction to THE MATHEMATICAL THEORY OF COMMUNICATION by Claude E. Shannon and Warren Weaver. Copyright 1949 by The University of Illinois Press. Reprinted by permission of the publisher.

"Summary of Principles for Explaining a Process" from BASIC TECHNICAL WRITING, FOURTH EDITION by Herman M. Weisman, p. 87. Copyright © 1980 by Bell & Howell Company. Reprinted by permission.

**Library of Congress Cataloging in Publication Data**

Arthur, Richard, 1946-
   The engineer's guide to better communication.

   (Scott, Foresman PROCOM series)
   Includes bibliographical references and index.
   1. Communication of technical information.
I. Reichert, Volkmar. II. Title. III. Series.
T10.5.A78 1984      620'.0068      84-1238
ISBN 0-673-15554-4

Copyright © 1984 Scott, Foresman and Company.
All Rights Reserved.
Printed in the United States of America

1 2 3 4 5 6-MAL-89 88 87 86 85 84

# CONTENTS

**FOREWORD** ix

**PREFACE** xi

**CHAPTER 1**

## Overview of Communication for Engineers  1

    The Engineer as Communicator  *1*
    Common Barriers to Communication  *6*
    Unique Barriers for Technologists  *9*
    Concepts of Communication  *13*

**CHAPTER 2**

## General Principles of Technical Communication  19

    Specific Audience Characteristics  *21*
    Describing and Explaining Concepts  *24*
    Technical Descriptions  *28*
    Involving and Persuading Audiences  *31*
    Visual Aids and Graphics  *32*
    Organizing for Effectiveness  *38*
    Summary  *42*

**CHAPTER 3**

## Oral Reports  44

    Manuscripts  *45*
    Memorization  *47*

Extemporaneous Speaking   47
Impromptu Speaking   51
Delivery   54
Conclusion   55

## CHAPTER 4

## Written Communication   57

General Suggestions   58
Written Style   61
Technical Style   63
Parts of a Formal Technical Report   66
Memos   76
Preparing Memos   78

## CHAPTER 5

## Gaining and Sharing Information   82

Interviewing   83
Developing and Structuring Questions   84
Communicating During the Interview   85
Concluding and Following Up on Interviews   99
Listening   99
Sharing Information in Groups   101
Beginning Group Interaction and Problem Solving   102
Application: The Reflective Thinking Format   105
Conflict   108

## CHAPTER 6

## Communication in the Technological Organization   111

Importance of Communication in the Technological Organization   111
Interfacing with the Public   113
Interfacing with the Technological Environment   115
Interfacing within the Technological Organization   116
Communication Load   119
Summary   121

*INDEX*   123

# FOREWORD

This volume is part of a series entitled *ProCom* (Professional Communication), which has been created to bring the very latest thinking about human communication to the attention of working professionals. Busy professionals rarely have time for theoretical writings on communication oriented toward general readers, and the books in the ProCom series have been designed to provide the information they need. This volume and the others in the series focus on what communication scholars have learned recently that might prove useful to professionals, how certain principles of interaction can be applied in concrete situations, and what difference the latest thoughts about communication can make in the lives and careers of professionals.

Most professionals want to improve their communication skills in the context of their unique professional callings. They don't want pie-in-the-sky solutions divorced from the reality of their jobs. And, because they are professionals, they typically distrust uninformed advice offered by uninformed advisors, no matter how well intentioned the advice and the advisors might be.

The books in this series have been carefully adapted to the needs and special circumstances of modern professionals. For example, it becomes obvious that the skills needed by a nurse when communicating with the family of a terminally ill patient will differ markedly from those demanded of an attorney when coaxing crucial testimony out of a reluctant witness. Furthermore, analyzing the nurse's or attorney's experiences will hardly help an engineer explain a new bridge's stress fractures to state legislators, a military officer motivate a group of especially dispirited recruits, or a police officer calm a vicious domestic disturbance. All these situations require a special kind of professional with a special kind of professional training. It is ProCom's intention to supplement that training in the area of communication skills.

Each of the authors of the ProCom volumes has extensively taught, written about, and listened to professionals in his or her area. In addition, the books have profited from the services of area consultants—working professionals who have practical experience with the special kinds of communication problems that confront their co-workers. The authors and the area consultants have collaborated to provide solutions to these vexing problems.

We, the editors of the series, believe that ProCom will treat you well. We believe that you will find no theory-for-the-sake-of-theory here. We believe that you will find a sense of expertise. We believe that you will find the content of the ProCom volumes to be specific rather than general, concrete rather than abstract, applied rather than theoretical. We believe that you will find the examples interesting, the information appropriate, and the applications useful. We believe that you will find the ProCom volumes helpful whether you read them on your own or use them in a workshop. We know that ProCom has brought together the most informed authors and the best analysis and advice possible. We ask you to add your own professional goals and practical experiences so that your human communication holds all the warmth that makes it human and all the clarity that makes it communication.

Roderick P. Hart<br>
University of Texas at Austin

Ronald L. Applbaum<br>
Pan American University

# PREFACE

Engineers face some unique challenges as communicators. In the explosion of technological advances, the task of communicating about technology becomes both more difficult *and* more essential. Often the traditional training of engineers provides only minimal instruction in communication. THE ENGINEER'S GUIDE TO BETTER COMMUNICATION is designed to assist those who work in technologically-oriented environments and must communicate both with other engineers and with nontechnical groups and individuals.

Other books on professional communication focus the majority of their material on technical writing. You'll find plenty on technical writing in this book, but there is considerably more. Since the greatest part of our communication is oral communication, a substantial part of this book deals with this vital form. Not only does it offer suggestions on how to structure formal presentations (such as speeches), it also offers valuable guidelines on managing the innumerable informal situations engineers must master to advance in their organizations. Structuring communication to develop and maintain effective and cooperative relationships, interviewing, and problem solving in groups, as well as other vital areas, are covered.

In addition, there is a chapter on managing communication—understanding and developing communication systems throughout your technological organization, across boundaries to other technological resources, and to the general public.

No one book can provide specific rules on how to deal with every unique communication situation or problem. This book *does* offer solid principles in a broad array of contexts. From these principles, practicing engineers—individuals who are, after all, skilled in applying principles to specific situations—can develop appropriate and effective communication skills which will serve them throughout their professional lives. This book contains numerous examples and exercises to help you understand how the principles can be applied.

This book could not have been written without the assistance of many individuals—many more than space allows me to acknowledge. Among those who didn't laugh when I mentioned engineers and communication are several who deserve special credit. First is Frank Reichert, who gave me numerous ideas about communication among engineers. Over the years, all my graduate teaching assistants have been a great help. Several, through their considered and considerable feedback about the effectiveness of my lectures and concepts, and through their own innovations, have especially helped shape this book: Bob Weissenborn, Jean Kebis, Rita Hoffmann, and Sally Henzl. I also wish to thank the many former students who have returned, not just to tell me how my course helped, but also to offer valuable suggestions as to what I could do to improve my efforts, based on their experience as working engineers.

Special thanks are due to Rod Hart and Ron Applbaum, for their well thought-out suggestions on framing and elaborating this book, and for their appropriate and understanding pressure. While any flaws which remain are my fault, credit is also due to Darcie Sanders of Scott, Foresman and Company, for her copy editing.

Finally, a most special note of gratitude to my family: to Jane, for tolerating and smoothing the disruptions of this project; and to Erin and Lauren, because they are who they are, and that's wonderful.

As I sit in my office surrounded by the many benefits of technology, it occurs to me that so many of them are taken for granted. The dedicated work and skills (including communication) that created these benefits are often little appreciated and less understood. I hope that this book can help engineers—not ony to facilitate their work, but also to develop greater understanding and sharing of perspectives among all of us concerned with, and affected by, technology and its applications.

Richard Arthur

# The Engineer's Guide to Better Communication

# CHAPTER 1

# Overview of Communication for Engineers

We all hear how there is a greater need for communication to help solve the problems of our society. Nearly every profession has recognized that its members have a growing need to communicate. The engineering sciences are no exception. Most engineers realize that communication is essential for survival in their field. Often, however, the need is not for *more* communication. In fact, most of us are barraged with communication—many times, too much to comprehend. The need is more for improvement in the *quality* of communication than in the *quantity*. This book is designed to help you refine your skills and improve the quality of your communication and, thereby, the quality of the organization of which you are a part.

## THE ENGINEER AS A COMMUNICATOR

Picture Carl Jarrett, an engineer and manager with the XYZ Corporation in California. Early in the morning, Carl is picked up by the car pool to drive to the office. On the way in, he and the other two engineers exchange complaints about the top management of the company and make plans for a family-oriented outing the following weekend.

Carl arrives at the office and, after exchanging pleasantries with the office staff, logs into the computer and checks his electronic mail. Three memos await him. One of them contains old news, since the author called

Carl at home the evening before. The second memo contains a request for information that Carl decides can wait until that afternoon. The third message, from the boss, was probably called in from a home terminal last night. It requires an immediate response. Carl quickly checks some specifications with an engineer in his section. Then, he types a response to the boss on the terminal, not waiting for a secretary to process his message.

After an hour of nearly uninterrupted work at his desk, Carl goes down the hall to the conference room for a staff meeting. During the meeting, the boss (who by now has read Carl's response) asks Carl to describe the progress on the current design project. Carl responds, "Just fine," but quickly notes from the boss's expression that more information is being requested. Carl proceeds with a six-minute description of the project to date. Another engineer suggests a solution to a problem plaguing the project.

Several other staff members begin to argue the merits of the suggestion just offered. Carl tries to evaluate each, but finds it difficult to sort out the reasoning each offers in the midst of freewheeling debate. The boss clears his throat, and in the immediate silence says, "This one's Carl's baby. Let's get on with other things." Carl notes on his pad that he will have to get back to at least two of the staff about their ideas.

Carl returns to the section he heads and calls together his staff. He tells them that the boss has made it clear that they are all "going to sink or swim alone" on this project, and that if they are going to keep their heads up they had better solve the problem. "Maybe a few extra hours this week might help," Carl says. He calls a staff meeting for that afternoon to work on the problem.

Skipping lunch, Carl responds to the message he'd put on hold earlier that morning. He gives the draft of the message to the secretary and tells her to get it out right away, and to be sure that the word processor is cleared for the report he will complete at home that night.

Carl goes to his staff meeting and attempts to control and facilitate the communication among his staff. Midway through the meeting, he leaves for another meeting downstairs, which will be with the sales rep and a potential customer who wants specific technical information. On his way, Carl stops by his office to get his jacket, only to have the phone ring. Carl decides to ignore it and goes downstairs.

The customer is accompanied by his general manager. Clearly, the manager knows his stuff. He asks specific technical questions. On the other hand, the customer, who is the owner of the company, is clearly not a technologist, but wants to understand what he is about to invest in. Carl starts to explain to the owner what he's just told the manager, but the sales rep interrupts, saying jokingly, "You'll never understand these engineer types. They speak another language. I'll give it to you in our lingo."

Carl returns to his office and checks for messages. Two more, requiring phone calls, are waiting. After completing his calls, Carl steps over to the coffee machine for a much needed jolt and becomes embroiled in a discussion of the Angels' current problems and what should be done with Reggie Jackson. Sally, a member of Carl's staff, comes up with what she thinks may be a solution to the design problem, which the staff split up to work on individually. Grabbing yet another engineer, Carl sits down, talks through the idea, and approves a test.

Carl notes that it's carpool time and debates whether to stay around and get a ride with Sally. Instead, he decides to carpool home, so that he can finish his feasibility report that evening. In the car, Carl is quiet all the way home, thinking about the report, the problem, and the boss.

A typical day for an engineer? Perhaps not every day for every engineer, but typical enough in many ways. From start to finish Carl spent large parts of his time communicating. Moreover, he not only communicated, but made decisions about communicating—when to communicate, what to communicate, how to communicate. He interpreted communication from others, not only for his own understanding, but for retransmittal to others, altering and embellishing as he went. He attempted to adapt his communication to suit various audiences, and came face-to-face with stereotypes which many people hold about engineers as communicators.

For an engineer like Carl, the ability to communicate often becomes as important as technical expertise. Without communication, he could not function; with poor communication his task would be immeasurably more difficult. Like most people, Carl handled some of his communication well and some not so well. With awareness and effort, he could reduce the poor endeavors and maximize the effective ones.

Since you already have this book in your hands, we presume that you have some awareness of the importance of communication for the engineering sciences, and are motivated to improve your abilities. The fact that you are reading this indicates that you have some communication skills. In fact, you would not be a part of the engineering sciences if you did not possess some basic skills as a communicator. We don't want to beat the point into the ground, but there is growing evidence to suggest that just *basic* skills in communication no longer suffice for persons seeking to improve their capacities as professionals. Communication is rapidly becoming a prime determinant of job opportunities and professional advancement.

Several recent studies show that a large portion of an engineer's time is spent communicating.[1] The successful accomplishment of engineering tasks is inevitably related to the ability to communicate among fellow technologists and with the managers of technological organizations.

These communication activities consist primarily of informal oral exchange of information; however, a substantial portion of our communication time is also spent in reading and writing, as well as in formal presentations before audiences. This book focuses on speaking skills in both formal and informal settings, as well as on writing skills, which are particularly critical for engineers. In addition, we hope to provide some ideas about designing communication systems in technological organizations.

A recent survey of employers of engineers in Wisconsin sought to determine which basic skills (not just communication, but the skills associated with a general education) were considered to be crucial for job success.[2] In addition, the survey attempted to discover the employer's perceptions of their employee's abilities in these areas of basic knowledge.

The survey revealed that, for engineers (as for several other professional groups), oral communication skills were ranked equal to computational skills in importance when evaluating job candidates. Interpersonal skills—the ability to communicate effectively and to manage relationships with others in the organization—ranked above these two areas, and were second only to general "attitude" as a hiring criterion. Thus, it would appear that the doors open easier for persons with better communication ability.

Furthermore, the survey indicated that engineers' supervisors perceived a greater weakness in oral and written communication skills in their employees than did supervisors of other professional groups. It would appear that either engineers come to their jobs less prepared in these skills, or that there is an extremely high demand for communication skills in engineering positions.

Engineering employers also reported (more frequently than other employers) that their employees "suffer career detriment because of writing and oral communication weaknesses."[3] Finally, this same report indicated that communication skills take on even greater importance for engineers as they move into managerial positions in the organization.

The findings of this survey are not unique. Many studies have demonstrated that engineers, like most other professionals, spend large portions of their work lives communicating. Communication skills are deemed so important that a large number of companies strongly encourage their engineers to take communication courses offered in the community or by the companies themselves. Perhaps many of you are reading this book as a result of your organizations' encouragement to improve yourselves as communicators. Engineers cannot hope to be particularly effective in the complex world of technology if they cannot share ideas effectively with colleagues. Cooperation is essential, and cooperation requires effective sharing of information through communication.

More and more, engineers are being called upon to become public persons, representing their organizations before an increasingly aware and concerned population. Businesses today expect employees to help them establish better public relations through the sharing of information. Technical organizations, while often vitally concerned with security of their ideas and processes, nonetheless recognize that an uninformed public can be a suspicious and distrustful force that, out of false fears borne of ignorance, thwarts the efforts of technology. Of course, not all the fears and mistrust are products of ignorance, and technological firms have also learned the importance of providing opportunities for their employees to listen to—as well as speak to—concerned citizens.

Engineers are frequently called upon to explain, if not to justify, projects to gain approval and support of legislators and of the taxpayers who frequently foot the bill for engineering projects. Many of these citizens are acutely sensitive to the impact of technology on the environment, whatever the source of funding. For example, environmental issues require open hearings and complicated and detailed written reports for scrutiny by government agencies, as well as members of the public.

In Milwaukee, Wisconsin, a massive project is underway to improve the sewage disposal capacities of the urban area. Engineers from this project have often spent large portions of their time not only developing the plans for this project (which will cost taxpayers over a billion dollars), but also informing, negotiating with, and ultimately persuading various interest groups that the plans are sound. The range is from formal presentations to technically sophisticated audiences to non-technologically experienced listeners to very personal conversations with one or two concerned individuals. Employees of this project could scarcely avoid becoming public representatives of the organization whenever and wherever anyone became aware of their tie to the project.

One engineer previously employed by the sewerage district, is currently employed with a local municipality. Here he finds that he cannot yet escape the sewer projects, since he regularly has to deal with citizens concerned by streets blocked because of construction, noisy equipment, and other problems associated with a large construction project.

In almost any engineering capacity, there is an increasing chance that you will have to communicate with the public. An inability to do so effectively jeopardizes you and your company.

Within the technical organization there is a growing demand for communication abilities. As technologists become increasingly specialized, the ease of communication among work group members decreases drastically. The knowledge gap between engineers specializing in different areas can be nearly as large as the gap between engineering and other fields, such as business or the social sciences. At the same time organizations require greater cooperation among a number of such narrow-range specialists, all of whom are necessary for completion of sophisticated and complex projects.

Ultimately, as well, most engineers are managers, and management of other workers can only be accomplished through communication. Individual skills, as well as some sense of the role and structure of communication in modern technological organizations, are essential for effective control and coordination.

Thus far, we hope that we have heightened your sensitivity to the importance of communication as a survival skill for engineers. However, recognizing the importance isn't very useful unless we can also recognize some of the difficulties associated with fulfilling this critical aspect of the engineering profession.

## COMMON BARRIERS TO COMMUNICATION

Engineers are faced with many difficulties in communicating. Some of these difficulties or barriers are faced by all communicators, regardless of profession; other barriers are specific problems confronted by engineers.

Several false assumptions lead to beliefs on the part of communicators that can hinder effectiveness. Below are a few barriers that we feel are often particularly harmful in communication. Several of these are adapted from Zelko and Dance.[4]

1. One prevalent belief has been that communication will take care of itself in time. This myth arises from the notion that we all know how to communicate if we try, so we needn't be too worried about improving. Communication problems, however, don't just go away most of the time. Instead, they tend to grow and compound their negative effects. We need to work at our problems if we are to be sure of overcoming them.
2. A second erroneous belief is that the ability to communicate effectively is something with which we are either born or not; that nothing can help us if we weren't "born with a silver spoon" in our mouths that guarantees our excellence in communication. The fact is that with work, anyone can improve their abilities. We suppose that genetics might have something to do with our capacity to communicate, but we firmly believe that a desire to improve offsets any inherent advantage of others. If we want to, and if we work at it enough, we can all improve. Those who appear to have always had greater ability simply received greater experience. Very little research even suggests that it is ever too late to develop abilities. If we—the author, editors and publishers of this book—didn't believe that people could learn to communicate better, then there would be no sense to our developing this material.
3. The third problematic assumption about communication is that there is one right way to communicate. The extension of this myth is the notion that there is some basic formula which, when mastered, guarantees effective communication in all situations.

These notions simply fail to recognize the inherent complexity of communication. While they may be useful concepts for those who would have you purchase some "guaranteed" system of communication improvement, they stand in the way of true development of communicative competence.

Quite simply, what works in one situation for one person won't necessarily work in other situations for another person. Communication is a creative process which reflects the individuality of the communicator. When we try to apply rigid formulas to our communication, it often sounds unnatural and forced. There are often *many* ways to communicate ideas effectively. This depends on the personalities of the sender and the receiver, as well as the situation and the goal of communication. Just as there are many ways to build a bridge that spans a given distance, so too are there many ways to span the distance between communicators. While the use of a mass-produced model "bridge" may be useful and functional in many applications, the mass-produced model is often ill-suited for the specific application. It rarely lends to recognition of the designer as a creative and dynamic talent.

Of course, the discussion of this myth is not intended to imply that there are no guidelines for communication. Just as in building a bridge, there are many factors that must be taken into account when communicating. Ignoring these factors may seriously reduce the utility of the communication. Nonetheless, we wish it to be clear that we believe that effective communication implies the necessity for unique, thoughtful "design" weighing and adaptation to any communication situation.

4. There is also an assumption that "All one needs to do to communicate effectively is to tell people what they need to know." There's really nothing wrong with this attitude, as far as it goes. Unfortunately, it leaves us short of our goal. Simply "telling" people what they need to know is no guarantee that they will either hear, understand, or respond to our messages. We need to be sure that we tell people in ways which (a) gain sufficient attention to be perceived, (b) can be understood from the receivers unique frame of reference as a communicator, and (c) are adapted to that communicator in such a way as to encourage the desired response.

We need to work hard to adapt our communication to others. What's clear to us is often very confusing to others. Moreover, we are all continually barraged with communication from various sources. We need to recognize that in the midst of this barrage a message that tells us what we need to know is often indistinguishable from what we don't need to know.

In many ways this myth has a ring to it which we believe lies at the base of many communication problems. This is the idea that we, as senders of information, somehow know what other people "need to hear." The reality is that this kind of attitude leads us to communicate what *we think* others need to know, not what *they think* they need to know. Effective communication necessitates our soliciting from those around us their perceptions of their own communication needs. We will almost always encounter difficulty if we try to decide for others what they need.

5. The final common barrier to effective communication is something that we often hear from students when they are required to take a communication course—they offer us statements that boil down to the attitude that "Communication is something that others, usually specialists, will take care of for us when we are employed in engineering." They point out that their secretaries will do their letters and memos for them, while technical writers will produce their reports. Even ignoring the naiveté which assumes that all engineers will have their own secretaries and access to technical writers, there are many problems with this attitude.

Specialists in communication need the raw information from which they will produce polished reports, etc. The engineer who has no notion of what comprises effective communication is usually unable to provide the raw material in a manner that assures ultimate quality. In addition, the engineer who leaves the communication to others risks severe problems. If you don't know what an effective letter, report, or oral presentation of your ideas looks or sounds like, then you have no way of controlling the quality of the material that goes out under your name. We contend that much of your communication is too important for your career to allow others to have total control. Most engineers wouldn't risk implementation of any project for which they were responsible if they didn't first assure themselves of the quality of the project. We believe that communication should be viewed in the same light.

Even if someone else produces all or part of your communication output, it is essential that you evaluate that material and correct it. Unfortunately, secretaries who combine office skills with the creative capacities necessary for effective technical communication are hard to find. Those who have those capacities now seek employment at levels where they can exercise their creativity—for example, as managers and engineers themselves. Technical writers, no matter how skilled, do not usually have the ability to judge the technical aspects of their writing; you must review their material to be sure that it accuately communicates your concepts to your audience in a way which reflects positively on you.

## UNIQUE BARRIERS FOR TECHNOLOGISTS

In our experience, technologists, by the nature of their education, personal orientations, types of jobs they perform, and the types of communication settings they encounter, often have some special problems which interfere with their ability to communicate effectively. Many of these are extensions of the common barriers we identified in the preceding section.

Perhaps the greatest problem for engineers is an attitude that seems entirely plausible: Facts speak for themselves. Technologists are trained to deal with scientific facts, and often presume that these facts are self-evident to others with whom they communicate. Engineers are trained in specific methodologies which enable them, by accepted standards, to determine the facts. Unfortunately, a great majority of others with whom engineers communicate have not had the same background or experience. That which is self-evident by scientific standards is unclear or incomprehensible to the general public. In addition, the standards by which scientists and technologists judge facts are often disputed (even if understood) by others.

Facts "speak" only if the listener or reader shares the same assumptions about what constitutes "reality." However, the problem in communicating facts is not always in having others understand the scientific basis. Often, the problem lies in interpretation of the implications of these facts. Non-technologists are sometimes inclined to weigh certain facts more heavily than engineers. When communicating, this mismatch frequently leads to frustrations on the part of both parties.

The real task for the technologist, then, is to help others understand—in ways fitted to their level of knowledge, understanding, and mental ability—why certain data from observation and research really are facts, which they should accept. When this is accomplished, the next essential task is to communicate—effectively—the interpretation and importance of these facts in decision making. Without these two steps, it is unlikely that even the most scientifically rigorous "facts" are going to "speak" to any audience with much impact. Facts don't speak, people speak facts—and they are only as credible as the receiver perceives them to be.

More often than not, there are also competing facts, which tend to either contradict or modify the interpretation of "our" facts. In a complex technological world, there are often conflicting explanations, even among those who are trained in the same fields. Even if there is broad agreement that certain specific data constitute facts, there is often dispute over how representative the available data are for the phenomenon under investigation. Many a decision is stalemated by competing facts of apparently equal validity. All the facts speak for themselves; unfortunately, they, say different things. We often become so convinced of our interpretations that we block communication which could help us satisfactorily resolve contradictions. Later in this book we will discuss conflict management techniques that are essential for solving these types of problems.

The final problem technologists have with facts is a result of their tendency to make decisions based on only the technological facts. That is, the facts considered relevant to decisions are restricted to those which are derived from narrow technological considerations of the problem area. The tendency is to ignore facts relating to broader aspects of technology; for instance, the human factors, which may be far more important than technological feasibility or desirability.

The novelist Neville Shute, in *No Highway*, describes a researcher who is so wrapped up in the necessity for pure technological research on aircraft that he loses sight of the implications for passenger safety. In his quest for technological knowledge, he simply fails to recognize that the value of decisions related to his projects must be weighed against human needs. The consequences of this failure would be disastrous, but for the intervention of another technologist, one who weighs the "nonscientific" factors in his decisions. Of course, in this novel everything turns out all right; the fatal flaw is detected and corrected before a tragedy, and everyone lives happily ever after. In real life, technologists sometimes fail to recognize the validity of factors which have nothing to do with the "pure" feasibility of their technology.

In our own community, the restructuring of the municipal sewage treatment facilities has taught many technologists how to address facts that go beyond technical feasibility. The success of their project will, in large part, be determined by their ability to listen to these other "facts," and creatively arrive at decisions which incorporate the broader perspectives of the population as a whole. Specifically, the facts relevant to decisions must also include those which address social and fiscal issues. When communicating to advocate technological programs, these other factors, of importance to the audience, must be addressed

Yet another problem for technologists as communicators is a tendency to approach the study of communication as a "hard science." This is really an intensification of the problems associated with the third common barrier to communication discussed earlier: The notion that there is one right way to communicate, with specific rules to follow, like a formula, in order to be effective. It is natural for an engineer to seek understanding of communication in this way. In fact, a great deal that communication scholars know about communication today is the result of the systematic study of communication conducted by technologists over the years. Unfortunately, however, there is much about communication which, as yet, cannot be easily understood from these approaches.

A large part of communication involves creative decisions, which make communication as much an art as a science. Engineers tend to think of products or processes that are somewhat absolute. There are beginnings and endings to the processes, and there are concrete products produced, often repetitively. Once factors are determined and weighed, the product or process can be repeated continuously by following the design (though there are usually some adaptations). With communication there are simply too many unique demands for each communication setting to successfully apply a universal formula. Moreover, there are too many variables in the process of communication researchers either haven't gotten around to investigating systematically, or which, as yet, aren't easily measured and quantified. Even the most dedicated scientific researchers of communication would not presume to offer a rigid description of the communication process, much less rigid prescriptions about how one should communicate.

Engineering students often ask us to provide them with a simple explanation of the important things to consider in their communication—the key factors which must be taken into account. To them, and to you, the reader of this book, we can only offer suggestions from our own experience and study of the area. You must however, recognize that these suggestions cannot replace your own creative adaptations of your communication. We offer no absolutes; nothing that works all of the time, nothing that never works anytime.

Often the very nature of the concepts engineers must communicate to others creates difficulties. Technological concepts have become increasingly complex and specific. The gap between engineers and the public, or even other engineers has widened. As in most professions, there is a great deal of "jargon" in the engineering sciences. This jargon consists of terms, or words, which, while easily understood by our fellow workers, either have no meaning or a different meaning to non–specialists. Frequently, we fail to realize that there are other possible interpretations for our words.

At a recent workshop, participants were given the task of describing geometric figures to one another. For one of the figures, several persons readily saw a similarity to the coffee cups we were using that morning. They explained the figure as a coffee cup, and were surprised to discover that several persons had drawn coffee cups, complete with handles, such as one finds in a restaurant or at home, rather than the styrofoam variety they thought they were cleverly describing. When we fail to account for the different possible meanings of words we are likely to have ineffective communication, even when the meanings appear—to us as the "senders" of communication—very concrete. Those with whom we communicate are often unaware of different meanings for the jargon. Even when the possible difference in usage is identified, many people are afraid to exhibit their ignorance and ask for definitions of such terms.

One of the ways that engineering concepts may be communicated involves the simplification of the overall idea or object. We all simplify our ideas when we communicate with children. Unfortunately, for engineers there are often times when simplifying a concept results in gross distortion. There may not be any way to make the idea understandable without leaving out critical subtleties, or underlying theoretical considerations. Engineers, by their profession, must strive for precision. They cannot afford to explain much of their work in imprecise, simplified fashions. On the other hand, the precision of our communication is useless if we are unable to obtain understanding from our listeners. Engineers must constantly evaluate audiences and balance the precision of their explanations against the possible needs and understanding of these listeners. Such a balance cannot be predicted by any simple formula, but again arises from the engineers' conscious creative efforts to communicate well.

One way to overcome the barrier between the specialized understanding of the engineer and the non-technical understanding of the audience is through the use of analogies (discussed in greater detail in Chapter 2). Analogies attempt to link complex and unfamiliar ideas, concepts, or objects, to familiar ones. Basically, they begin with a comparison of the complex object with the familiar object. In the previous example of the coffee cup, the persons describing the truncated cone did so by comparing that object, for which most of them did not know a specific name, with a familiar and supposedly easily understood object. While analogies are rarely exact comparisons, and therefore result in a degree of distortion, they certainly allow for some understanding. Unfortunately, many technological concepts are so specialized and complex that no analogies to the experiences and knowledge of the average listener exist. When this situation occurs, the possibility for distrust and conflict between technologists and non-technologists is likely to increase. Of course, we are not suggesting that you, as an individual, educate your "public" to a sufficient level of technological expertise to be able to understand your concepts. We do hope, however, that you grow to recognize those barriers which exist, and try to develop sufficient dialogue to promote the understanding you need. If you give in to the feeling that the barrier of technical/non-technical understanding is too great to surmount, and therefore not worth the effort, the results of such thinking may spell professional disaster for you and for your organization.

We feel that it is important to recognize that many of the people with whom you may have to communicate not only *can't* understand your technology easily, but may be *unwilling* to do so. Robert Pirsig, in his book *Zen and the Art of Motorcycle Maintenance* (a book which we recommend to anyone concerned with the barriers to technological communication, as well as to those who enjoy a good book), addresses the specific barrier of the unwillingness of some individuals to cope with technology.[5] These people erect mental barriers and refuse to listen to technical explanations. They wish to leave the pure technology to the technologists; on the one hand reaping the benefits of that technology, while on the other hand controlling the negative effects. They do not realize that technology cannot be divorced from the uses to which it is put; become disappointed when technology is less than perfect, yet avoid contributing to improvement through understanding.

Alvin Toffler discusses similar ideas in his book, *Future Shock*.[6] Toffler discusses how technological (and social) changes occur so rapidly in modern society that many persons live in fear of these changes, and grow to oppose all development that alters the way in which they lead their lives. They react by opposing all change, regardless of its potential benefits, simply because the change represents a disruption of their current lifestyle. Some may even wish to turn back the clock to simpler times, when technology, as they believe, wasn't confusing their lives.

These resistances—the unwillingness to be informed or understand, and the hope that change can somehow be stopped or at least limited to their own vision of how things should be—are increased by the public's image of technologists as different and detached from the concerns of everyone else. When engineers (and other technologists) choose to isolate themselves from the public (perhaps because the communication barriers are so great), the public image of them is reinforced. Engineers, therefore, face the burden of initiating communication to overcome the fears of the public. Simply put, engineers cannot afford to hide out in their labs or offices or anywhere else, and ignore the public. They must learn to welcome the opportunity to communicate with the public as (at least) an open door; then work hard to overcome any misunderstanding through communication. The engineer must work with the non-technologist to assure technological growth in ways which reduce mistrust and uncertainty.

## *CONCEPTS OF COMMUNICATION*

We have described some common barriers to communication, as well as some specific difficulties engineers and others in the technological sciences face. We trust that we have developed a sense of the complexity of the task of improving communication. We believe, however, that the task is not *too* great. With attention and concern, communication for engineers may be improved substantially. To do so requires some basic understanding of the communication process. In describing some of the characteristics of communication, however, we recognize that there is imprecision in our analogies. We hope that you will remember that we are introducing only a few of the important ideas about how communication works.

*Figure 1-1  Shannon-Weaver model*

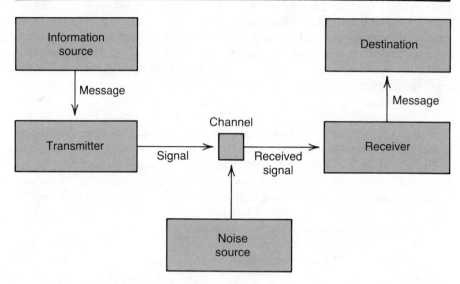

Perhaps one of the more useful concepts of communication was introduced by an engineer—Claude Shannon, who was the director of Bell Telephone's research laboratories in the 1940s. Shannon, along with Warren Weaver, developed a model of communication (Figure 1-1).[7]

This model not only recognizes that there are senders and receivers of messages, but that there are various channels through which messages are transmitted. Each channel has its specific capacity and capability. Certain channels carry specific information more efficiently than other channels. All channels are subject to *noise*, or interference with the message. This noise limits the transmission and reception of messages. As technologists, Shannon and Weaver were particularly concerned with reducing the noise in channels: eliminating static, fading, distortion, etc.

Noise has come to be an important concern for communicators. Beyond the *physical noise* that can limit channels of communciation, we also need to be concerned with *psychological noise*, the interference each individual creates in the communication process. We've already given several examples of psychological noise: future shock reactions which reduce our willingness to listen, confusion over the meaning of words, and other barriers to communication. In a sense, then, what we offer you by introducing the term noise is a convenient label, a piece of jargon, which we hope can simplify our discussion of communication throughout the remainder of the book. Our task as communicators amounts, in many ways, to the identification and elimination of potential sources of noise in our communication with others.

Communication necessarily involves the actions and responses of at least two people. Each of us has our own wealth of experience that leads us to send and receive communication in ways that are different from any other person. When we fail to recognize our differences from others, the potential for noise increases dramatically.

David Berlo identifies some further elements in the communication process which we feel are useful, particularly as they help us in the identification and elimination of noise. Berlo developed a model (see Figure 1-2) which not only describes the general process of communication, but also some of the characteristics of the persons and messages involved in communication that influence communication effectiveness.[8]

In the next chapter we will discuss the potential influence of source and receiver characteristics. Then we'll move on to describe strategies for the selection of appropriate content, treatment, and codes for messages based upon the individuals communicating.

As listeners, as well as senders, we are actively processing the messages we receive. We don't remember what we hear or read word for word. Few, if any of you, could remember, word for word, what the last sentence contained without referring back to it. Most of you, however, could remember the basic idea of the sentence. Noise often enters into our

*Figure 1-2  Berlo model*

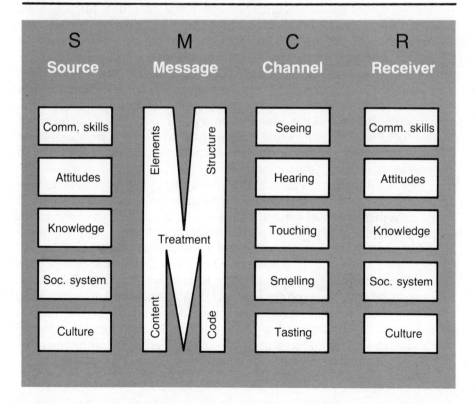

communication when we take what we hear or read and put it into our own words for memory. We are generally quite effective at this process, but the possibility always exists that we may choose to remember the wrong things and forget the important ones, or that we distort the information as we recode it. Effective communication involves adjusting our messages to help listeners remember the important information in a way that matches our intended meanings. Effective listening involves checking to be sure that we have captured these meanings properly, rather than assuming understanding.

The words, or symbols, which we use as a large part of our communication are often imprecise. We have already discussed jargon and misunderstood coffee cups. Even such a simple word as "run" has over one hundred definitions in many dictionaries. When we communicate we must make sure which definition applies. Noise enters when we assume this, rather than take time to be sure.

We will be effective in reducing noise only when we recognize that each situation, each individual, and each channel is unique and interacting in a unique fashion. This complex uniqueness makes a complex prediction situation and is the reason why we cannot offer you a simple formula for guaranteed communication effectiveness. This interaction of factors is even more complicated, because communication is a process wherein each of these variables constantly shifts and develops over time. Just when we begin to feel confident that we have achieved the proper wavelength, we discover that something new has been added to our communication. Therefore, we can only suggest that you become more aware of your communication and develop a wide range of communication skills. This strategy will allow you to adapt to the various and variable communication demands made upon you. We hope that the remaining chapters will assist you in widening your skills, as well as help you to understand the potential sources of noise. We believe that what we have to offer you represent some reasonable starting points for the communication process as encountered by engineers in their professional lives.

In the following chapters of this book, we will attempt to introduce you to some key concepts of communication for engineers. Chapter 2 contains a discussion of general principles of technical communication. These principles apply to the specific contexts and forms of communication described in later chapters. Chapter 3 provides a description of how to prepare and deliver oral presentations. The discussion includes both formal and informal reporting. Chapter 4 discusses the critical forms of writing that engineers often face. Among these are letters, memos, abstracts, and formal technical reports. In Chapter 5 we present information about structuring your communication for interviewing (information seeking), problem solving in small groups, and maintaining solid relationships with your colleagues. The final chapter sets forth some concepts on how the overall communication system of a technical organization can be structured and managed.

## Questions and Activities

1. Keep a log of your communication activities for several days. How much of your average day involves some form of communication? What form does that communication take? Which communication activities did you feel you handled well, and which ones caused you problems?
2. How many of the problem areas, or barriers to communication, were attitudes which you yourself held to some degree?
3. Think about your professional goals. As you move toward those goals, how might your communication demands change? What skills will become more important? How might you work to develop those skills in advance of increased demand for their use?
4. How do "facts" as understood by technologists differ from a general audience's conceptualizations of fact?
5. How has increased specialization in technology increased the problems you encounter as a technologist? How could you help to overcome these personal problems?
6. How often of late have you encountered someone who wishes "technology would just go away and leave me alone?" What do you think causes people to hold such an attitude? How might you, as an engineer, through your communication, help overcome such an attitude?
7. What are five common sources of noise you encounter in your communication?
8. Develop a list of receiver characteristics, based on Berlo's Model, which you think may inhibit the ability to understand communication of technical information. Draw up a list for different types of audiences: fellow workers, non-technologists, management, yourself, etc. Try to think of ways in which you might adapt your communication to match these characteristics.

## References

1. Erickson, H., "English skills among technicians in industry." *Communication and the technical man*, T. Wirkus and H. Erickson. Englewood Cliffs, New Jersey: Prentice Hall, 1972.
2. Page, P., & Perelman, S. *Basic skills and employment: An employer survey.* Madison, Wisconsin: Interagency Basic Skills Project, University of Wisconsin System, 1980.
3. Ibid.
4. Zelko, H., & Dance, F.E.X. *Business and professional speech communication.* New York: Holt, Rinehart and Winston, 1978.

5. Pirsig, R. *Zen and the art of motorcycle maintenance.* New York:Bantam Books, 1974.
6. Toffler, A. *Future shock.* New York: Random House, 1970.
7. Shannon, C. & Weaver, W. *The mathematical theory of communication.* Urbana, Illinois: University of Illinois Press, 1949.
8. Berlo, D. K. *The process of communication.* New York: Holt, Rinehart and Winston, 1960.

# CHAPTER 2

# General Principles of Technical Communication

In the previous chapter we identified some of the reasons why engineers and other technologists need to be concerned about their communication. We also identified some of the barriers to effective communication that engineers are likely to encounter in their professional lives. In this chapter we will introduce some guidelines to improve technical communication. We caution you, again, that these are only *general* guidelines, not absolute specifications to be followed to the letter in every communication setting. We hope to share with you some of our insights and those of other persons concerned with this area, so that you can begin to develop your own set of guidelines for future communication.

We believe that the *critical starting point for effective communication is developing an understanding of your audience.* This belief is not restricted to technical communication, or the professional communication of engineers, but applies as well to our efforts to communicate with other professions, and in our personal lives with our friends and loved ones. To be successful as a communicator, you need to recognize that the people with whom you communicate are unique human beings. Each of these unique persons requires unique messages in terms of content, your selection of treatment, and code.[1] While it may often be possible, or necessary, to adapt our communication to large groupings of individuals, the extent to which we can allow for specific individual characteristics within a larger plan will, in part, determine whether we will be accurately understood, and, ultimately, whether our ideas and proposals will be accepted.

What are some of the factors that may require you to adapt your communication? Berlo's Model in Chapter 1 gives us some broad categories to consider. Berlo suggests that both senders and receivers of communication share certain key characteristics that determine their receptivity to, and understanding of, communication. Each of us is, to some degree, a product of our respective culture and social contacts. The social environment that we have been a part of may heavily influence the ideas we have been exposed to and the experiences we have gathered. Each of us, on any given topic of conversation, has some varying amount of previous understanding or knowledge. We are all more or less skilled as communicators; and therefore more or less able to listen effectively, to manage conversation with others, and to produce messages that are accepted and understood. Perhaps most importantly (and often as a result of the other factors in Berlo's Model), we each have unique combinations of feelings or attitudes about various concepts or topics all of which we communicate. All of these may create noise between communicators.

As we prepare to communicate, then, it seems that it may be wise to try to determine some of these factors about our potential audience. One of the common starting points for the analysis of an audience is the identification of *demographic factors*. These factors are generally considered to be the variables of their social and cultural background and current status. They include such things as age, gender, education, income, profession and experience. Not *all* of these variables will be likely to influence how we communicate in *all* settings, or across *all* topics of conversation. Each may, however, be relevant in the situation in which you must communicate. Therefore, we believe that it is often worthwhile to ask yourself if any of these factors may be likely to cause your listeners, or the persons you are listening to, to react in some unique way: to have different meanings for words, to be likely to listen to or remember different things, or to consider some things to be more important than others.

In the process of assessing the demographic characteristics of any communicators, it is essential that you recognize that these factors may provide only a very loose prediction of individuals. While in some sense it may be helpful to draw a broad picture of others based upon their age or sex or any other factor, we must realize that there are often as many differences *between* members of groups as *between the groups*. Assessing demographic characteristics and then making predictions about communication on the information may lead to very inaccurate stereotypes about communicators. These stereotypes may lead us to offend the persons to whom we are trying to adapt. Nonetheless, this information may help us in developing ways to link our ideas to our audience. For instance, speech teachers are frequently asked to identify effective public speakers for students to use as models for their own speaking. It does the students little good if we identify Martin Luther King, Jr., or Jack or Robert Kennedy, if the students were born in the early 1960s and have little likelihood of having ever heard a speech by these individuals.

Similarly, it would do little good to describe a technical process by comparing it to crop rotation if the audience is composed of city folk. Examples only work if they relate to the experiences of the listeners. Demographics may provide clues to help us determine the likely *range* of experiences our audiences may have. The demographics will rarely predict the specific experiences of each audience member.

Younger audiences may be more concerned about environmental issues, while older audiences may be more concerned with the impact of the technology on tax rates and inflation. The business manager may be most concerned with bottom line profit projections, while the engineer may be most interested in the creative technology employed in a project. Women may have a greater interest in how your ideas influence or affect women. Highly educated persons may become bored with simplified explanations. All of these factors may be true at any given time, but human behavior is so complex that we recommend you never allow yourself to totally rely on such broad sterotypes based upon demographics, unless you have adequately studied your audience and verified your assumptions.

Our final words on demographics are about gender differences. The engineering sciences have traditionally been dominated by men (at least in numbers). As a result, much communication by engineers refers exclusively to males in its examples and in the choice of specific language. Whether you agree or disagree with changes in the status of women in your field and elsewhere, these changes have occurred. It seems wise, for whatever reasons, to recognize women through the form of your communication. It can only court disaster to fail to recognize that women are likely parts of your audiences and that they expect recognition. Broad stereotypes, which assume that women are either uninterested or unable to understand the specific technology, will inevitably lead to anger and poor communication in most modern communication settings. You can rarely afford this hostility, regardless of your beliefs about the "proper place of women." (A notion we feel inhibits communication with anyone.)

The specific educational background of your listeners or readers is far more likely to predict their degree of openness and understanding than their gender. While we personally find the use of phrases such as "he/she" or "his/hers" to be somewhat awkward, we recognize that these may be more effective once we become accustomed to them. Another option is to use plurals, which are gender-neutral, whenever possible. At the very least, when we refer to persons in our communications we can alternate between references to either sex, and include examples of women as well as men.

## SPECIFIC AUDIENCE CHARACTERISTICS

Besides the broad demographic factors, there are other characteristics about your audience that may create noise if you don't recognize and adapt to them. One of these factors is the specific knowledge your listener or reader has about your subject area. No one enjoys long presentations that repeat information they already understand. We also don't enjoy long presentations of information that is totally new to us. We become "turned off" to communication that is either too unfamiliar or too familiar.

Dr. Frank E. X. Dance, a noted communication scholar, notes that too little research has been satisfactorily completed to give us an idea of the balance between that which is known and that which is unknown by the audience.[2] From his own experience, and that of others, he concludes that a starting point in balancing presentations may be in the range of 70 percent familiar material to 30 percent unfamiliar. Obviously, this specific balance is not always possible, or desirable. Intelligent and motivated audiences possessing general familiarity with a given topic area may respond better to a 50/50 balance. Audiences low in listening motivation, or desiring only summarization of ideas, may require much higher proportions of familiar material.

The point is, you can't begin to adjust the amount of new information unless you have some idea of what your audience already knows. Demographics such as age, experience, and education may help you to determine the likely level of audience knowledge, but active discussion with members of future audiences should be sought to help verify your assumptions. With highly complex technical data it is likely that your audiences will tire rapidly, so the goals you set for yourself as a communicator may have to be reduced, based upon the amount of previous knowledge of your receivers.

All of us have limits to our ability to process information. When we are overloaded, we may elect to stop listening, or we may begin to seriously distort what we hear. We decrease our accuracy by boiling down the information we receive into the personal summaries we remember. We referred to this type of noise in Chapter 1. You can't provide a crash course in everything your audience needs to know to become experts in any given area. Provide them with the information that is critical to approach an understanding at a level that they need to make reasonable decisions.

Sometimes it makes us feel like great experts to be able to mystify our audiences with complicated facts and explanations. However, all that information may not help the audience understand and accept your ideas, since it exceeds their level of knowledge. Audiences aren't impressed with your expertise if you haven't communicated it to them in a useful way. For instance, it is not usually necessary to tell every listener *how* your computer works, if they only need to know *what* the computer is capable of doing, and at what cost to them. You must try to determine, in advance, not only the knowledge your audience already has, but also what they need as a result of your report.

Besides the determination of knowledge in the specific topic area of your communication, it may be useful to analyze your audience to find other areas of knowledge. This may, again, be useful in deciding the types of examples you can use to help explain your ideas. Try to choose examples from the areas in which your audience has the greatest background knowledge, especially if you yourself have some understanding in this area.

Probably the most critical audience factor to consider when you are attempting to gain acceptance of your ideas is the attitude your audience brings to the communication setting. In Chapter 1 we talked about future shock. If your audience is generally afraid of technological change, then it will be essential for you to overcome this barrier before they will be likely to listen to the merits of your proposals. An audience that already supports your plans in principle doesn't need a long explanation of why they ought to believe what they already accept. They need inducement to do something about what they believe. You may adjust the content of your message accordingly.

An audience which is initially opposed to your proposal needs to be shown why your ideas are sound and desirable. They may be hostile, defensive, and resent your avoidance of those issues that they feel are the most important. An *immediate* plea to them to run right out and support your proposal will likely increase their resistance and make subsequent arguments ineffective. Audiences which oppose your position need to be moved slowly and carefully to your position.

Research by many scholars suggests that we usually hold positions on issues that are not just a choice between "for or against."[3] Rather, we have a range of positions which we feel are acceptable, and a range of positions which we consider unacceptable. The speaker (or writer) who merely identifies audience attitudes as either "for" or "against", and sees the issue as one which only has these two alternatives, is unlikely to persuade an opposing audience.

Effective persuasion with opposed audiences requires, as a minimum, a realization of the possible "rightness," at least in part, of the opposition. Strong advocacy of positions on the issue that fall too far from the audience's range of acceptable responses will only reduce the likelihood of successful persuasion.

An example may help to clarify this. If you were attempting to persuade an audience to accept the development of a new nuclear facility, you might determine that one part of the audience favors such a plan, and another part opposes it. If you could separate these two parts of the audience, then the first half—those who favor the facility—might only need some brief reinforcement of their belief and instructions about what they should do to help assure that the facility will be built.

The other half of the audience would be unlikely to respond favorably to your plea to go out and take action in support of the facility. The same speech or report wouldn't work for them. Here you would need not only the knowledge that they were opposed to you, but also of the positions that they might accept. They may, based on their advance understanding, oppose your facility, but be willing to accept it if they believed that adequate safeguards were built in to protect the community. Your knowledge of this specific concern can help you to structure your message to reduce their objections about safety. It may well be that you can only open the door for further communication through your own persuasive effort. Many efforts to persuade audiences fail because they attempt to move the audience too far too fast. Merely getting the audience to believe that the design engineers really *are* concerned about the safety of the community may not win the audience over to support, but it may make them more willing to listen to other arguments later. We may then shift the audience attitude in gradual steps over a period of time.

We have identified several factors that we feel may be particularly important to you in planning your technical communication, regardless of the channel or purpose of that communication. We believe that there may be as many important factors about your audience as there are factors in the personality of humans. In any communication setting, you need to begin with an analysis of your potential audience. Try to anticipate the variables that are likely to create noise and resistance in the audience. As you begin to plan your message, you may then try to overcome these problems, rather than blindly attempting to communicate with everyone in the same fashion. Some specific techniques matched to your audience may help you communicate effectively.

## DESCRIBING AND EXPLAINING CONCEPTS

As we've mentioned, technical communication involves, as often as not, complicated ideas. These ideas may be very hard to get across to audiences because we cannot easily simplify them. In an earlier example we discussed the description of an object: a "coffee cup." Unfortunately, our example was an instance where the description didn't work, but it did represent a good starting point for technical description.

### Analogies

Most of us manage to make sense out of the things we encounter in our lives by comparing them to other things. When we do so we draw an *analogy* to something in our past experience. Analogies are one of the most powerful tools in the engineer's arsenal of technical and communication weapons. Engineers are in the business of taking ideas from the "pure" areas of science and placing them in the realm of practical application. That is, the engineer draws an analogy from one application of a concept to another problem area and arrives at a solution. Engineers who can readily identify the relationships of one application to other novel situations succeed as technologists.

Similarly, engineers who can recognize the application of one concept from their technical experience, and extend that to the realm of experience of non-technologists, succeed as technical communicators. We have, ourselves, just developed an analogy that compares the effectiveness of technological problem solvers with the effectiveness of communicators. As in most analogies, the comparison isn't perfect. The ability to develop analogies isn't the only thing that engineers must do in order to succeed in technology and communication, but it is a starting point in both places.

Some of the groups faced with the task of describing our poor old coffee cup went about the process by describing each and every line, every angle, every dimension in relationship to one another. This partitioning of the figure is another effective way to providing technical descriptions, one which we will discuss later in the chapter. This method is useful, but often time consuming. The analogy, or the use of concrete examples, can help us shortcut the process of description, but only if we are sure that the comparison is made to something that the receivers recognize and understand. Also, the comparison should be to something that has only a limited amount of ambiguity, or possible interpretations, itself.

A student once attempted to compare the operation of a nuclear fusion plant to the functioning of the sun. That would have been a good analogy, except that the receiver didn't have any better idea of how the sun works than of the fusion plant. We all use analogies constantly as we communicate, but we rarely take the time to think about them and plan them with a specific audience in mind. When we do, not only can we communicate our ideas so that they are understood, but in a way that is creative and pleasing to the receiver. Consider the following analogy:

### Pulse Laser Light

Pulse Laser light is different from ordinary light in the same way a marching group of people is different from a disorganized mob. A mob moves randomly, moving in all directions at the same time. This is like the motion of ordinary light, where the particles of light, called photons, move in all directions at the same time. In a marching group, the people move in the same direction. Laser light is like this, all its photons move in the same direction at the same time. In a mob, all the people move at different speeds, which is like the different frequencies in ordinary light. The different frequencies, or colors of light, are randomly distributed, and vibrate at different speeds. Laser light is all one color, vibrating at the same speed. The people in a marching group all move at the same speed.

Not only do they move at the same speed, but the people moving in a marching group also move in step. Each person moves his right foot at the same time as all the others, and does the same with his left. Laser light is like that also, all its photons move in phase, will all the photons reaching the peak of their vibrations at the same time. The lowest point, like the point in a step where the legs are side by side, is also reached at the same time. In ordinary light, even if the light was monochrome light, all one color, and sent out in a spotlight beam, it would not yet be laser light, as the photons would not be in phase. This can be explained as even if a mob were to voluntarily move in one direction, at the same time, at the same speed, it would not be a marching group. The marching group has a precision that comes from its discipline and training, that a mob would never have.

Laser light is more concentrated and intense than ordinary light, because all its photons deliver their energy at the same time, in the same way a marching group will arrive, all together, at its destination.[4]

---

You can see how the concept of a laser not only is clarified (especially for someone who isn't an engineer), but, becomes alive as well, in a pleasing and novel fashion. Our bet is that this student will do well both as an engineer and as a communicator, perhaps because the two are so interrelated. If they learn to use the talents they have developed in one aspect of their profession in other aspects as well, engineers really do have the advantage over most other professionals as communicators.

A couple of suggestions about using analogies are in order. First, be sure that you don't overstate the comparison and distort your concepts. While it might be useful, for instance, to compare humans to computers as information processors, there are some important differences between the two. These differences are as important as the similarities if we wish to understand either of the two. The comparison provides an anchor against which we may also draw contrasts to help the receiver. If we neglect the contrast as a part of the full elaboration of our analogies, we have lost half of the value, as well as risked oversimplification and misunderstanding. Thus, while the comparison of the brain and a computer is useful to identify certain concrete operations which both may conduct, it helps us to understand the computer better if we also understand its limitations for abstract and creative thought, or differences such as short term and long term memory processes.

Second, summarizing what we mentioned earlier, be sure that the supposedly familiar part of the analogy is really familiar to your audience. A few pages back, we gave the example of a student who compared nuclear fusion with the sun. This was, for his audience, comparing two unfamiliar concepts. The analogy failed to assist the process of understanding.

A third axiom is that you should avoid making the comparison more complicated than the original object or concept. An analogy that attempts to describe a simple piston compressor by comparing it to an automobile engine may be on the right track, since an automobile engine is familiar; but the engine is also far more complicated than the compressor. As another example, a colleague recently had a conversation with an electronics engineer who described what was needed to repair a malfunctioning computer terminal. The engineer described the needed part as "like a lot of transistors." Courtesy prevented the non-technologist from responding, "So what?", since the analogy created only a glimmer of understanding. Again, two equally unfamiliar and complicated articles were compared.

As an aside, it's worth noting here that the colleague who chose not to reveal that he really didn't understand the analogy was doing what most listeners tend to do. He was unwilling to display his ignorance in the face of a description that implied that only an idiot wouldn't know what—some thirty years after their appearance—a transistor was. Listeners will rarely let you know whether your analogy has been understood properly. Ego interferes. You must monitor your analogies and their results very carefully. The consequences of the author not understanding the analogy were minimal here, but could lead, quite literally, to disaster elsewhere. Check with your listeners. See if they understand. Don't assume anything. It would have been very simple for the engineer to ask whether the listener knew what transistors are, and what they do.

A final rule for use of analogies is to avoid stretching a comparison too far just for the sake of having an analogy. Analogies won't work if two concepts really have only one little thing in common, while differing in most ways. Comparing a catalytic converter on a car to a scrubber on a smokestack, even assuming that the listener is familiar with the point of comparison, still might not work if the only thing that the two have in common is that they remove emissions. If they go about this process in a completely different way, the comparison becomes highly misleading; unless, of course, it is sufficiently qualified as to the limits of the comparison. That limiting of the comparison may take more time and energy than the benefit of the comparison justifies.

## *Examples*

A second useful device for technical communication is the use of examples. As with analogies, an example will only be useful if it represents something with which the receiver is familiar. For instance, if you wished to describe an overload protection device, it might not be very effective to give an example such as the circuit breaker at a power station. On the other hand, most people would be familiar with the circuit breakers or fuses in their own homes. The examples you use should be easily visualized by your audience. That, after all, is the purpose of examples: to give your receivers something concrete, which will give them a mental image of whatever you are describing.

Use examples and analogies whenever you are trying to communicate ideas that are not easily understood by your audience. Work on developing a file of analogies and examples for the concepts, the objects, and processes you typically must communicate to others. Don't leave this to chance. Our own experience is that these devices are very difficult to think up when we are on the spot and need them in a hurry. When you think of or hear a good analogy or example related to your area, write it down and save it. Effective communicators develop these files of ideas and useful concepts to ease the task of preparing reports at a later date.

## *TECHNICAL DESCRIPTIONS*

We've already discussed some of the problems and some of the methods used in describing objects or concepts in technical communication. The difficulty of describing objects to an audience unfamiliar with them may be overcome through the use of analogies and familiar examples. However, there are other situations in which these devices may not be sufficient to assure adequate description.

Describing an object from technology involves answering several key questions in the minds of your receivers. Weisman suggests several of these questions: What is the object? What does it do? How does it do what it does? What is it made of? What are its parts? How are these parts related to each other in the objects as a whole?[5]

These questions don't necessarily represent every possible question in the minds of receivers, nor is each always present. They represent what we feel are reasonable starting points for the development of descriptions. If you can answer these questions for your audience, the likelihood is that they will have at least some understanding of the object. When combined with the use of visual aids or graphics, these answers are particularly powerful. We will discuss the use of graphics later in this chapter.

A general outline of a technical description usually begins with the definition of the object. The definition may be the specific technical description, but it is unlikely that this will aid many receivers in understanding the object. It does us little good to define a rabbit as "Oryctolagus cuniculus" for someone, unless their background is strong in Latin or Zoology. Definitions only help to achieve understanding when the components of the definition are already familiar to the receiver.

Weisman recommends that formal technical descriptions follow the general definition or description, with a division of the object into each of its main parts (assuming that there are sub-assemblies or parts). Each part is described individually through the use of examples, analogies, or, if these aren't available, by describing the dimensions, angles, appearance, composition, etc. Each part is then described in relationship to other parts in the final assembly, including where and how the attachment is accomplished. The first step in the description usually involves taking the object through a complete cycle of its operation: indicating input, internal action, and output. An example would be the description of a power station. The input of coal, gas, or other fuel begins the description. The characteristics of the fuel are detailed for the reader. The combustion process and apparatus are described next. Following this, the movement of steam through the turbine is discussed, including description of the turbine characteristics. Operation of the generator comes next. Finally, the output of electricity and steam is detailed. Condensing of the steam and recycling finishes the initial description. A summary of the major components of the operation reinforces the process description for the receiver—fuel input to combustion to steam to turbine, generator to electricity output and recirculation of condensed steam.

You might also organize a description according to the order of assembly. This is especially useful if you have the object and can go through the assembly with the audience, or can develop graphics and overlays (transparencies representing the separate parts) which show the step-by-step assembly. Thus, you might describe a pump by first providing specifications of the left-hand case. After this, the left-hand bearing assembly would be discussed, followed by the impeller assembly, the right-hand bearing assembly, and the left hand casing.

Building on this, or as a separate description where necessary, you can describe the cycle of operation as well—showing how the pump works, including input internal action and output, as in the object description outlined above.

Technical process descriptions really build upon the object description. The process description contains many of the same critical components: who uses the process, why it is used, where it is used, and the special requirements necessary for the process to work efficiently. The description includes explanations of the objects: the tools or machines that perform each step in the process. Thus, object description is always a part of the process description. Your goal is to describe what is being done and how this is accomplished in sufficient detail for your audience to comprehend and (ideally) visualize the whole operation.

Consider a description of the process of municipal solid waste resource recovery systems. You could begin with the description of the characteristics of the solid waste: what its composition is and what components are considered recoverable. The waste enters the system and metals are separated by electromagnets. Following this, the remaining material moves to a trommel (can you think of a suitable analogy to describe a trommel to an audience?), where further heavy material and glass is broken up and separated. The remaining material moves by conveyor to a shredder. The shredded material enters an air classifier, where a high pressure air stream separates lighter, more combustible material. The air stream carries the light material through ducts to a cyclone, where the material is collected for use as fuel.

This description of waste recovery is missing several essential details. Each of the separate processes and pieces of equipment should be discussed. The use of graphics for the whole process, and of each specific step in the process, would assist immeasurably in assuring comprehension by the audience. Try to think of analogies and examples that would work for describing the above process. The air classifier might be compared to the process of winnowing—separating the wheat from the chaff—which many audiences might already be able to visualize. The fully elaborated description of the process of solid waste recovery should also include some detailing of the efficiency of the process, where it can be used appropriately, who uses it currently and with what results, and potential problems in the process. Not all of these elements will be necessary all of the time, but you should ask yourself as you prepare your description whether the audience will require or be interested in these or other facts.

An old rule of writing for newspapers suggests that each article should answer the questions: *Who? What? Where? When? How?*, and *Why?* We believe that most descriptions would benefit from analysis and attention to these questions. For technical descriptions, the *How?* comprises the most important element, but the others ought not be entirely neglected.

Your descriptions of objects and processes must be adapted to the needs, abilities, and knowledge of your audience. Following an appropriate organizational pattern will help understanding, but even well organized descriptions often fail. Work on examples that fit your audience, then fit them into a clear organizational structure. Avoid a constant shifting back and forth between steps in the process, or between objects or components. Treatment, or organization, is critical. Finish each step in your description before moving to the next step.

### Summary of Principles in Explaining the Process[6]

The explanation of a process, especially one that involves the fabrication of a product, or the conduction of an experiment or activity which leads to a specific result, should include the following information:
1. The definition of the process.
2. A description of the time, setting, performers, equipment, and preparations.
3. An indication of the principles behind the operation.
4. The listing of the major steps in chronological order.
5. A step-by-step account of every action and, if appropriate, inclusion of description of apparatus, materials, and special conditions.
6. Details under major steps, arranged in chronological order.
7. A concluding section, which may be a summary or an evaluation of the process.
8. Drawings to aid in the explanation of crucial steps and in the vividness of the description of an action.

## INVOLVING AND PERSUADING AUDIENCES

In Chapter 1 we mentioned that audiences often respond on the basis of their distorted beliefs about technology. Unless you are able to overcome their initial resistance by getting them interested, and by establishing your credibility to them, it is unlikely that you will achieve positive results from your communication.

Creative and vivid analogies, particularly those which assist the receiver in connecting technology to their own personal experiences, can assist you in gaining the attention of your audience.

Once attention is gained, you must be concerned with the credibility—the "believability"—of your ideas for your specific audience. To do so, we frequently use authorities in a given area as support. This is very effective, but only if your audience is somehow willing to accept that the authorities you cite are, in fact, experts in the area. It does little good to indicate to an audience of non-technologists that a specific person they have never heard of recommends a specific process or concept. If your audience won't recognize the authority, you must prove to them that he or she is a credible expert. State what the person's qualifications as an authority are.

Engineers often face the difficulty of persuading groups to accept some type of development which may have an impact on the environment. It does little good only to cite authorities from the ranks of engineers who support these ideas. It would be far more effective if you could identify members of groups who are usually identified with opposing environmentally questionable developments, but who support your current plan. To use a contemporary example, audiences fearful of technological advances would probably not be impressed if you used as an authority James Watt, the former Secretary of the Interior who developed a reputation for favoring technology. On the other hand, an officer of the Sierra Club who favors your proposal, would be a much stronger supporting authority. Support from the Sierra Club might "turn-off" some business person. When President Reagan nominated Sandra O'Connor for the Supreme Court, support from conservative politicians did nothing compared to the impact of the endorsement by the liberal Senator Edward Kennedy. Once a key supporting figure who might have been expected to oppose the nomination was identified, the issue was never in question.

Avoid presenting your ideas in ways which make you appear self-serving. Audiences often suspect that engineers advocating technology are only trying to fulfill their own needs, not the general needs of the public. In our city, for instance, a group attempted to oppose the plan for sewer modifications. Later, it was alleged that this group opposed the current plan only because they, as contractors, wouldn't get a piece of the pie. The plan they favored would have increased their chances of gaining contracts. Whether their plan would have been better than the original no longer was the issue. Their credibility was in doubt, and their arguments lost all impact.

We must argue in our own interests on many occasions, however. When we do so, it may be useful to recognize that our position is not the only possible one. We need to present balanced arguments that exhibit to the audience our objectivity and impartiality to the greatest extent possible. Attacks on opponents to our proposals may be personally satisfying, but they are rarely effective. The starting point for effective persuasion is the acknowledgement of opposing perspectives. Then comes a clear, concise description of why the position we favor is *superior* to other positions—not the *only* possible answer.

Nothing could be more disastrous than for an engineer to imply to an audience, either through a statement or an attitude, that, "You don't know enough about engineering to understand whether we should do this." The audience expects you to present information that will help them make reasonable decisions. Their response is likely to be that you don't know enough *about people* to make decisions. Remember, since it is you who are attempting to persuade the audience, it is you who has the obligation to prove to the audience that you are correct. They don't have to discredit you or your ideas; you must determine the ways in which you can make yourself and your ideas credible for your audience. Don't insult your audience, or imply that you are somehow more qualified *in all aspects* than they.

Finally, try to link what you advocate to the beliefs of the audience. By this we mean that you should attempt to show your audience how what you are proposing actually fits what they are already in favor of. For instance, if you can show your audience that your plan is more efficient, or less costly, or protects rather than harms the environment, (when these are the primary concerns of your receivers), you will be far more likely to succeed than if you merely indicate that the proposal fits only *your* idea about what is important. What *you* believe isn't as important to them as what *they* believe. It is your obligation to show them that what you both believe really isn't too different. The use of ten solid, logical, arguments in favor of something often fails if there is an eleventh issue of primary importance to your audience that you have neglected to address.

## VISUAL AIDS AND GRAPHICS

A large body of research indicates that receivers process information in different ways, depending on the channel (oral, visual, etc.) by which it is communicated.[7] Using as many channels of communication as possible helps to assure understanding and impact. When channels are used together they can both supplement and reinforce one another. Ideas that simply cannot be explained through spoken words come alive when they are made visual. Pictures that are meaningless by themselves can be easily interpreted with a few clear words of explanation. The human being is equipped with a number of senses, and when they are all brought into play in communication (though not necessarily all at the same time) the increase in understanding can be dramatic.

Visual aids or graphic displays particularly help us because they provide a way in which many ideas or components can be combined. The words that we *speak or write* to describe things must be expressed sequentially, so it is often difficult for receivers to develop a clear idea of how everything combines. Our memory for discrete information, especially information presented to us orally, is often limited. A particular case of this is the ability to comprehend and remember large numbers or lists of data. The visual aid can help us concentrate our descriptions into one package with instant impact, rather than a sequential effect. The old adage states that a "picture is worth a thousand words." However, the picture may be useless if we don't provide our receivers with the words they need to understand. Visual aids and words necessarily go together for maximum effect.

Engineers all have some basic understanding of graphics, and many are quite good at developing effective aids for readers and listeners. Where they seem to have some difficulty is in determining where to use graphics, and for what purposes. Sometimes engineers tend to use their visual aids without providing the words and explanations needed to make them understood. They assume that the picture alone is sufficient to explain an object or a process. As a general rule, aids or graphics serve best as *supplements* to the written or spoken descriptions, not as the *primary* (or only) means of description.

To be effective the aids must be clear to your receiver. A visual aid for an oral presentation that is too small is worse than useless, since it distracts and irritates the audience. So too for visual aids which are cluttered with so many details and labels that the receiver is overloaded with sensory information. You must keep aids simple, especially when they accompany oral reports. Include only the information and labels which are essential to your receiver's understanding your presentation. Eliminate the unnecessary.

Graphics which accompany written reports can contain greater detail, but still must maintain legibility and a sense of open space within them. They must always be neat, or your audience will perceive that you are unconcerned with detail, and discount other important aspects of your communication. Stick figures and freehand drawings have little place in technical communication unless you are particularly talented and precise. The same care that goes into the preparation of specifications and plans on the job is needed in the preparation of aids and graphics.

Aids are particularly helpful when we attempt descriptions through the use of comparisons and contrasts. Two aids combined can readily show an audience how two objects or processes are similar, and how they are different. When we compare figures such as percentages, a graphic presentation of bar graphs makes comparisons much clearer and easier for an audience. It is usually easier for the audience to understand from such a graph that 46 percent is nearly half again 32 percent, than if we merely speak or write those figures. This is especially true if we have multiple comparisons of figures, which the audience would have difficulty remembering over the period of time necessary to read or listen to them all.

Remember that the purpose of the aids is to supplement other means of explanation. They are not a total substitute for analogies, examples, or detailed descriptions. Don't use aids just for the sake of having them as "eyewash." They should be included only as they assist you to make your point. If they aren't needed—because your audience already understands, or is familiar with the concept or object—eliminate them from your presentation.

When explaining processes it is often useful to develop aids which match the stages of your verbal description. One aid can depict the overall process and the combination of all the parts of that process. A series of aids may be used to depict each sub-part, or stage of the process. For process descriptions it is especially effective to develop transparent overlays, which allow you to present the process sequentially by adding each step to the earlier steps. This works both for spoken presentations with the use of an overhead projector, as well as in written reports.

Your aids create an overall impression of you as a communicator and as a professional. Companies often spend thousands of dollars for specialists to develop aids to support presentations by executives and sales personnel. Large productions are often computer assisted to coordinate a vast array of projectors and audio equipment. We mention this because these sophisticated presentations set a standard against which your efforts may be evaluated. Audiences grow to expect this wizardry, and are less and less impressed when presentations are accompanied by aids which aren't integrated into elaborate productions, no matter how carefully developed. On the other hand, the carefully coordinated presentations may serve to create a gap between you and your audience, since they involve careful timing, which may reduce your ability to adapt to the unique demands of your audience. The personal touch may be lost under these circumstances. While these multi-media wonders may have immediate impact and please audiences, if there is no substance underlying their usage, no consideration of how these aids all contribute to the goals of the presentation, then the impact will likely be short-lived. Included on the following pages are some examples of graphics.

Figure 2-1 is an example of a simple bar graph. Such bar graphs are useful for comparing several objects or processes across several criteria. Thus, it pictorially represents the relationship of three specific pieces of

*Figure 2-1  Comparison of three personal computers*

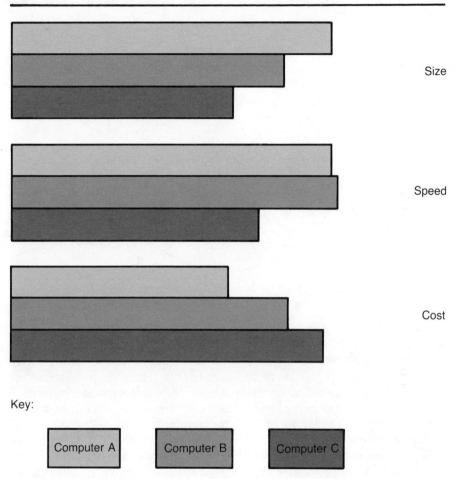

equipment on three categories of important characteristics. Labeling is clear; the comparisons are relatively easy and immediate.

Figure 2-2 is an example of a pie chart. Similar to the bar graph, the pie chart is useful for static comparisons of allocation, or distribution, but the pie chart is used for showing the distribution on only one criteria or category. The pie chart helps the reader sense the contribution of parts to the whole. Use of contrasting colors in separate sections would enhance the visual impact and increase clarity.

A third form of graphic is shown in Figure 2-3. This shows a line graph which is most useful for showing trends over time. An important consideration in constructing the line graph is in the scaling. Intervals and scaling within each axis must be equal to allow fair comparison. Thus, it would be inappropriate to shift the scale of units produced from an interval

**Figure 2-2  Average time spent on communication tasks by engineers**

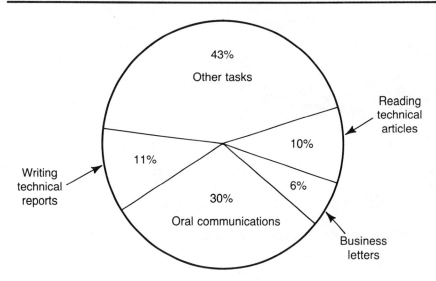

(Source for statistics: Erickson, H. "English Skills Among Technicians in Industry," in T. Wirkus and H. Erickson, *Communication and the Technical Man*. New Jersey: Prentice Hall, 1972.)

of 1,000 units to one of 2,000 as the absolute number increases. Figure 2-3 shows one factor over time.

The line graph, as shown in Figure 2-4 is also useful to show relative changes in time for two or more factors. This figure shows a relatively simple, straightforward comparison where the two factors clearly diverge. Often, however, the line of the factors may cross and recross making it difficult for the reader to follow the trends and comparisons. Use of different colored lines can help the reader or viewer. Note how, on Figure 2-4, the producers of this graphic have not clearly differentiated the lines for humans and robots. The average reader could infer this, perhaps, but nothing should be left to chance. Furthermore, clear labeling eases the receiver's task and increases impact.

Figure 2-5 is a graphic delineating a process. No attempt is made at actual representation of the physical characteristics of the components. This will usually be an adequate form for an audience familiar with the components; if not, separate graphics can be developed for each component. These components can be referred to sequentially in the discussion of the process. The form used in Figure 2-5 is especially effective in presenting a relatively uncluttered representation of the whole process. Remember that spoken and written language is necessarily linear—it cannot provide at one instant a representation of the whole. Thus readers or listeners are required to hold the components, as described, in short term memory in order to conceptualize the total process. The graphic simplifies what would otherwise be an almost impossible task for anyone who is not already familiar with a complex process. So many components

**Figure 2-3   Productivity before and after training programs at Plant X**

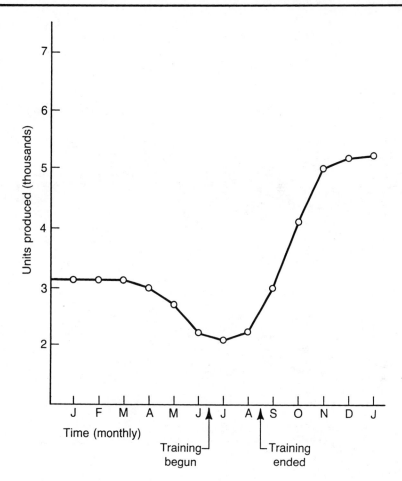

in relationship to one another are likely to be confused and perceptually misplaced if presented without a graphic for reference.

Figure 2-6 is similar to Figure 2-5, but provides a graphic representation of the components of a process. This is most useful when the audience is likely to be unfamiliar with the nature of the components. They would have difficulty following your description, due to their lack of the ability to visualize the parts, as well as the whole.

This particular form is only useful if the process involves a relatively low number of stages and components to be represented in this fashion. The example we have provided is already too crowded to be visually pleasing and easily comprehensible. We feel that the use of the type of graphic shown in Figure 2-5—with separate, and larger representations of the components—would have been far more effective in presenting this process.

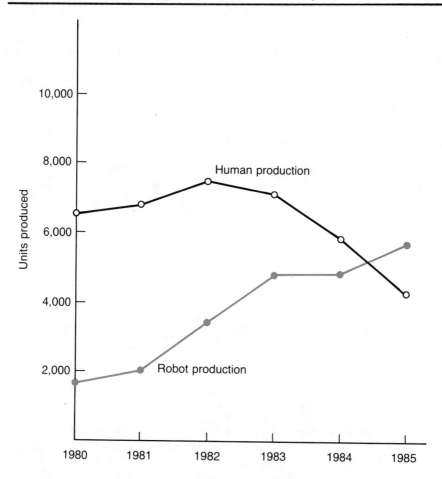

**Figure 2-4 Actual and projected human vs. robot production at Plant Y**

## ORGANIZING FOR EFFECTIVENESS

Berlo's Model, in Chapter 1, identified three key components of messages: content, treatment, and code. We have discussed some of the content that may be appropriate in technical presentations. The treatment—or organization—of that material is equally essential. We will discuss the code—the language you use—in later chapters.

The structure of presentations may seriously affect their impact. Your audience, depending on its level of motivation, may only be concerned enough to seek specific information, and not other parts of your report. They may be unwilling to listen hard and follow you through your presentation if it is unclear where you are leading them, and how you plan to get there. The most general pattern for most communication is to: tell

*Figure 2-5  Coal liquifaction process*

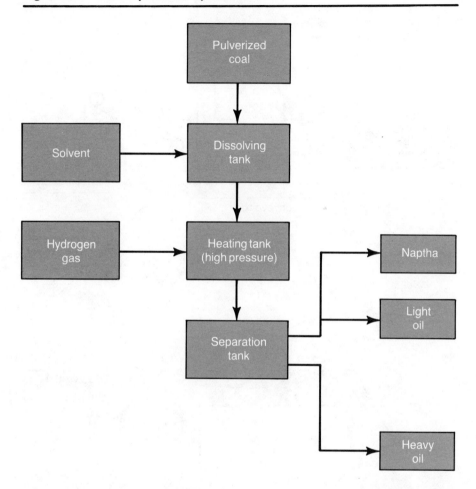

them what you are going to tell them, tell them, then tell them what you told them. You assist your audience by informing them of the purposes of your presentation and the basic pattern you will follow at the very outset. This applies not only to reports, but to even the information sharing interviews we all utilize on nearly a daily basis. Our receivers are better prepared for what follows if they are informed in advance.

Audiences have a tendency not to listen to everything, or to read every part of a report. We all daydream, or skim over what we don't consider to be important; especially if we are busy and have many other things on our minds. To assist the reader or the listener, it is helpful if we build into our communication some degree of *redundancy*, or repetition. We don't mean that everything in your communication needs to be repeated two or three times. We suggest, however, that you include brief summaries of your

**Figure 2-6   The process of producing RDF from MSW**

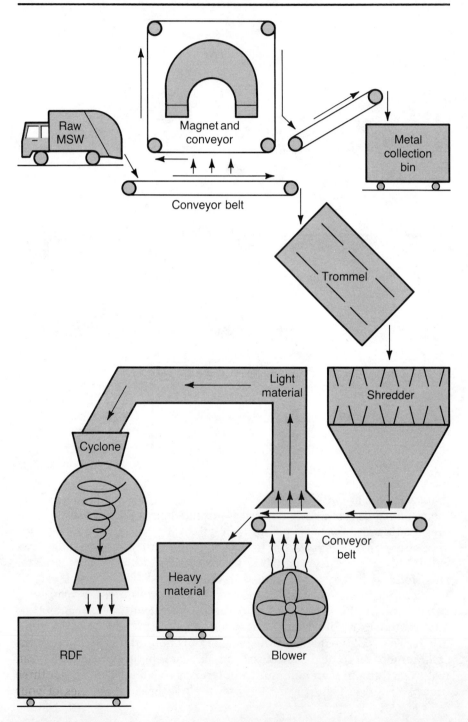

information as you move from point to point. For instance, as you complete the description of one component of an object or process, provide a summary of what the component is and does, then link that component to the next area, as well as to previously described components. To apply this to our own writing, we might now say to you, "We have already discussed the importance of audience analysis to help you determine the content of your presentations. We have also suggested some specific strategies such as analogies, examples, and graphics, which help the reader to visualize your verbal descriptions of concepts. We have indicated that organization is equally important, and that you should both preview your presentation and summarize your points to assist your receivers in understanding. Now we would like to suggest some specific organizational strategies for specific audiences."

When you summarize, it may be useful to include a new analogy, or example, or present the visual aid again, in order to reinforce your earlier discussion and give it new impact.

Reflect the needs of your audience in your organization. The boss may wish to know the "bottom line, up front." The boss may be primarily concerned with your proposals or recommendations, the costs, and the expected benefits or profits, and have little concern for the technical aspects of how you arrived at your decision.

Fellow engineers may wish to see a clear discussion of the methodology and assumptions prior to any discussion of recommendations. They may feel that the process through which the decisions are reached is more important than the decision itself.

The marketing manager may be concerned primarily with the sales implications of your recommendations, and wish to find out this information at the beginning. The personnel department may wish to know what the recommendations will require in allocation of workers. Each of these potential groups has different information requirements, and may be only slightly interested in other parts of your report. Where possible, adjust the organization of your report so that the information needed by specific readers is found easily, particularly in the introduction and the conclusion of your discussion. If you are making oral reports, technical briefings, etc., you may be able to adjust your introduction and the conclusion for different groups to which you make the presentation. For written reports it is more difficult to write several reports which highlight information for each reader. The recent addition of word processors to the business environment may make this type of adjustment far easier, and is worth your consideration. At a minimum, include clear headings for sections of written reports, which help the reader to locate pertinent sections.

## SUMMARY

To conclude this chapter on general principles we offer these suggestions:

1. Recognize each person that you communicate with as a unique individual with unique experiences, knowledge, and attitudes.
2. Try to match your choice of content and treatment to the needs, beliefs, and expectations of your receivers.
3. Use examples, analogies, and descriptions that not only help your receivers to visualize your ideas, but do so in a way which keeps them interested.
4. Use visual aids and graphics to *supplement* your ideas. Keep them simple and clear. Use several channels to reinforce your ideas, but don't overload your audience on all channels simultaneously.
5. Don't directly attack the integrity of those who oppose you. Recognize that it is your obligation to change them, not for them to defend their own beliefs.

### Questions and Activities

1. Think of the major communication events you have encountered in the last several weeks. What kinds of audiences did you encounter? Develop a profile of demographic and specific audience characteristics of a specific target audience. What difference might these characteristics have made in how you should have communicated technical information to them?
2. Choose an issue about which you feel particularly strongly. Make a list of the reasons, or arguments, you have for your position on the issue. Next, make a list of the reasons which someone with whom you disagree might have for their position. Try to develop arguments first *against your own position*, then against those who disagree.
3. Choose two or three objects or concepts with which you deal professionally on a regular basis. Develop an analogy for each of these that would help a non-technically trained person understand each. It will probably be helpful to envision a specific individual or audience. If possible, try out your analogy. Be sure that your analogies are sufficiently developed to really clarify. To say, "RAM is like tape recordings, and ROM is like a disc record," is a start, but probably needs greater elaboration for most audiences.
4. Describe a technical process, including description of the components and the entire cycle of operation.
5. Develop a visual aid that would enable an audience to envision the entire process described in (4) above.

## References

1. Berlo, D. K. *The process of communication.* New York: Holt, Rinehart and Winston, 1960.
2. Dance, F.E.X. unpublished lectures, 1976.
3. Sherif, C. W., Sherif, M., & Nebergall, R. E. *Attitude and attitude change: The social judgement-involvement approach.* Philadelphia: W. B. Saunders, 1965.
4. The author is indebted to Jeffery S. Jones for permission to use this sample.
5. Weisman, H. *Basic Technical Writing.* Columbus, Ohio: Charles E. Merrill, 1962.
6. *Ibid.* pp. 166-170.
7. McLuhan, M. *Understanding media.* New York: McGraw Hill, 1964.

# CHAPTER 3

# Oral Reports

We noted in Chapter 1 that oral skills comprise an important determinant of the effectiveness of engineers. In fact, the presence or absence of these skills may well be a factor in determining employment opportunities, as well as the chance to move up professionally. We cannot, in the limited space of this book, cover every possible oral skill. We suggest that you may wish to use the book in the *ProCom* series which specifically discusses oral communication. (*Professionally Speaking: A Concise Guide*, Robert J. Doolittle). For our part, we will provide a general discussion of some relevant decisions that technologists must make, and some appropriate skills which form a foundation for effective speaking.

At the outset, we wish to make it clear that we consider oral communication and, specifically, oral reporting, to go beyond formal scheduled speeches or even briefings. We believe that, in many ways, we make oral reports each time we discuss our jobs with other members of the organization of which we are a part. When the boss drops by and asks, "How are things going," our response is a report to the boss. We believe that it is possible to map some of the concepts of good formal speaking back onto our thinking about even these *seemingly* casual encounters.

Oral reporting, regardless of context, requires clarity of language and organization. Rambling statements, which meander from point to point full of useless words and ideas, dissipate our effectiveness. We also believe that good oral reporting requires flexibility on the part of the speaker. The more responsive and adaptable you are, the better you will be able to achieve your communication goals.

The strategies for technical communicators identified in the previous chapter are useful tools for oral reports. There are other factors about how we present our ideas that are critical.

When faced with formal, planned reports, we must make a decision about how we are going to prepare to speak to the audience. One decision is the choice of what is termed the *mode of delivery*. This is a fancy term for whether we choose to read our report, memorize it, or somehow present our ideas without a specific script. Generally, there are four modes of delivery which correspond to the choices above: manuscript, memorization, extemporaneous, and impromptu. The choice of mode depends on the skills of the speaker, the audience, and the type and complexity of the material to be presented.

## MANUSCRIPTS

Manuscripts generally allow us to present complex material in a very precise fashion. Manuscripts are scripts for the speech which contain, word for word, what we intend to say to the audience. Often they are read precisely as written; though many times, through practice, we become familiar enough with the material to present portions of the material without reading.

Manuscripts are especially useful when there are many facts and figures, which would be hard to accurately remember and state. Scripts also help us if there are a large number of quotes from authorities or other sources which must be presented exactly. Reading may also be helpful when we present material which has been developed by a group of individuals other than ourselves. Sometimes we could not possibly present their ideas so precisely, or as well, in our own words. Manuscripts are also appropriate when there is a need for a precise record of what was said, or when statements of policy are made for which the organization may subsequently be legally accountable.

However, for many speakers there are some serious drawbacks to reading manuscripts. First, most speakers are unpracticed at reading in a natural and interesting fashion. Most speakers who read tend to drone on in a monotone after a few minutes. Speakers who read seem to sound detached from their material, as though they themselves don't know, or care, what they are saying. Because they are reading, the audience often fails to feel that the speaker is working at the process, or is involved and cares about what is being said. Good readers, the "Walter Cronkites of the world," make what they read sound as though they are using their own spontaneous words. These people read material frequently. Most of us are only called upon rarely to give a prepared speech. For these reasons we generally recommend that speakers try to avoid the use of a strictly read manuscript.

There are, however, ways in which manuscripts may be adapted to help overcome these problems, and others as well. Particularly, it may help with what we believe is the greatest problem of all for technologists: manuscripts tend to "lock you in" to a predetermined course throughout the presentation. They tend to isolate the speaker from an ability to converse with the audience, to adapt to the audience's responses. These responses may take the form of outright questions, statements that the audience doesn't understand some part of the material, or such subtle non-vocal cues as furrowed eyebrows, slumped posture, stifled yawns, or the glazed eyeballs of a thoroughly confused listener. In the face of these signals, most manuscript readers seem to cruise blithely along through their script; either failing to observe the audience reactions because their eyes are glued to the text, or, deciding that, since the speech is all prepared, it will be presented that way come hell or high water. When we commit ideas to paper we seem to feel that they are carved in stone, and cannot be changed.

Good speaking requires that we accept that our best laid plans often fail us. If we use manuscripts, then we must be prepared to abandon—or at least supplement—the planned material when it fails to accomplish the goals we have set for our speech.

One manner of adaptation of the manuscript is the use of a half-page manuscript. This technique places the text to be read on the right-hand half of each page. On the left-hand side we place comments or suggestions related to the presentation of the material, such as a note to read a portion more slowly, or to pause before a particularly important point. We may note to ourselves that it is time to lift our noses and look at the audience, so that the audience will feel as though we really came to speak to them. Lack of eye contact may make the audience feel as though we could as well have sent a tape recording, for all the involvement and presence we exhibit. Consistent eye contact with the audience tends to enhance our credibility, and the audience's sense of our trustworthiness.

The left half may also contain some supplementary material, which can be incorporated into the speech if it is apparent that some of our ideas aren't sinking in and the audience seems lost. You may note some other examples or analogies you can use to further clarify your ideas, if necessary.

Regardless of the form of manuscript you use, if you are going to read your planned speech, be sure to practice it *out loud* prior to your presentation. We write language differently from the way we speak it. Often, when we hear ourselves speaking what we have written, it becomes apparent that certain phrases sound strained, or unclear, or, as is often the case, overly pompous and arcane.

## MEMORIZATION

The second mode of delivery is memorization. Unfortunately, for all but the very best speakers, memorization leads to dull, inflexible, distant presentations. Typically, memorized speeches lead to all of the difficulties of manuscripts (since they are manuscripts committed to memory), plus the additional disadvantage of struggling to recall the material without assistance. Because we are a society that is heavily dependent on the use of written language, we seem, most of us, to have never developed particularly good memories for large amounts of prose. Most of us struggle to remember the words to just a few short repetitive songs, in which we have the additional assistance of rhyme and rhythm. In our experience, we have seen many attempts at memorization end with the speaker either staring desperately at the ceiling in the hopes that the manuscript will miraculously appear there, or with seemingly endless pauses followed by repetition of the last part that the speaker does remember.

For the technologist, who is often involved in presenting material that involves complex formulas and figures, we can think of few—if any—advantages to memorization. We believe, however, that if one chooses to use a manuscript, that it is useful to be familiar with your script to the point where you can spend most of your time looking at your audience, rather than at the script. Learn portions of the text, but have the script available; both for reference to complex material, and for assistance if memory fails during the stress of the presentations.

## EXTEMPORANEOUS SPEAKING

Extemporaneous speaking involves the preparation of an outline of your speech and careful practice of the presentation, but leaves the final choice of words until the speech is underway. As a result, the speaker adapts the speech to achieve the purpose of gaining understanding, rather than the purpose of merely presenting the material.

Extemporaneous speaking is not "speaking off the cuff." To be effective, there are several stages of preparation. These stages of preparation are the same, up to a point, as those you would use in the preparation of a manuscript speech. We cover them here because we believe that the majority of your oral reporting will almost inevitably lend itself to extemporaneous speaking, especially in the somewhat common form of technical briefings.

First, you must determine exactly what your purpose is in speaking to your particular audience. Usually, your topic will have been determined by the situation or circumstances of your talk. That is, when we are asked to talk in an organizational setting we are rarely asked to "speak on anything you like." The topic is related to some professional function. Therefore, when we mention determining the specific purpose, we are referring to the process of narrowing the scope of your presentation within the general topic area. For each speaking occasion you should analyze your audience to determine their specific needs for knowledge within the topic area.

The specific purpose of a speech should be drawn up in such a way that it is only one sentence in length. If it is more than one sentence, then it is likely that you are identifying several subpurposes, rather than the specific integrating focus of your presentation. For instance, if the management of the organization asked you to "give us a short update on your project," your general topic area is determined for you. The specific purpose, of course, depends on the needs of that audience in relation to the topic, but might be something like, "To inform the managers of the division's progress and problems on Project Alpha during the last month." Two parts of that purpose differentiate the purpose from the topic. First, the potential speaker has determined that it is appropriate to discuss problems as well as progress. Second, the speaker has decided to limit the presentation to the last month (perhaps the period of time since the last progress report), rather than to the whole history of the project.

Perhaps we should note that our own feeling is that it isn't all that important how the purpose is worded, as long as *you* are clear in your own mind *exactly* what you wish to accomplish in your presentation as you begin your presentation. We believe that it is always worthwhile to stop before diving into the preparation of a report. Consciously decide what the speech is going to do. This clearly identified purpose then helps you to choose appropriate material to include in the presentation. It helps avoid extraneous material that really doesn't relate to your purpose for that speech. It is also worth noting that determining the purpose in this explicit, specific, fashion often forces you to really think about why the manager asked you to make the report or presentation. We feel that one of the keys to success may be the extent to which you attempt to match the purpose of your presentations to the underlying expectations of the audience. Ask yourself, "What do they really want to know, and what do they really need to know to do their jobs?" Match your purpose to your answer, to the extent that you are able. *We are not, however, saying that you should only tell people what they want to hear at all times.* We are saying that you may be able to be more effective in conveying the material you wish to others, whether it is good news or bad news, when it comes in a package that meets their perception of the focal area of the presentation. "Surprise" presentations of material that the audience is mentally unprepared to receive often fail to carry much impact, since the listeners are operating in a different *frame*. Their frame of reference is what enables them to understand your communication. When their frame is different than your frame, it is likely that there will be distorted communication. It's like when someone speaks up behind you and asks a question, and you can't understand it because you are thinking of other things.

Having determined your specific purpose, it is often useful to write that purpose at the head of a sheet of paper. Then begin to list the possible ideas related to that topic that you wish to cover. Zelko and Dance suggest that you next take separate pages of paper for your introduction, your body, and for the conclusion to your speech.[1] You then transfer ideas from the page with the general list of ideas to the specific parts of your speech where you feel they will be most effective. Which ideas will gain the attention of your audience, and help them to develop a strong sense of your purpose at the outset of the briefing? Which points will provide a succinct summarization of your purpose at the conclusion, or will leave your audience with a strong lasting impression of the importance of your material? Remember, as you ask these questions, to keep both your specific purpose and your knowledge of the audience in mind.

Once you have determined a broad division of material into the introduction, the body, and the conclusion, decide how each section should be organized. Choose some logical pattern to your ideas, especially in the body. Engineers often have to describe processes, so it may be useful to either divide the process spatially into its subparts, or chronologically according to the sequence of events in the process.

One of the most widely used and accepted organizational strategies for technical reporting is the problem-solution pattern. Engineers are frequently called upon to propose technological solutions to specific problems in their oral reports. To be effective, it is essential that you and the audience have some degree of agreement about the nature of the problem to be solved. The problem-solution pattern helps to assure that you and your audience have the same frame for the problem prior to the discussion of the solutions. If your audience has an image of the problem which differs from yours, or if they don't recognize that there is a problem to begin with, then there is little likelihood that your solutions will be accepted. Extemporaneous speaking is especially effective for this pattern, since, should your audience have any misunderstanding, or if you have misinterpreted your audience's concept of the problem, it allows flexibility to clarify the problem area. Extemporaneous speaking is more readily adaptable to spontaneous questions from the audience, as well as either vocal or nonvocal indications of a desire for greater clarification.

Once you have placed your ideas for the speech in a logical order, your next task is to identify ways to clarify and support your ideas. Determine examples and analogies to help your audience gain a sense of the concepts. Determine the specific authorities or statistical support which will help enhance the credibility of your ideas with the specific audience you will address. Determine where visual aids will be necessary to supplement what you will say, and begin the task of assembling or developing these aids.

At this point the preparation of an extemporaneous speech takes a different route from manuscripts. A manuscript would be developed literally, detailing exactly what you will say. The extemporaneous speech does not rely on a script. Therefore, your task at this point is to begin to practice your presentation. The actual words, or the *code*, which express the ideas outlined, will vary each time the presentation is given—in practice and in the actual speech. It may be useful, however, to have transitions written out, as well as key statements in the introduction or conclusion. Specific quotations and summaries of statistics, formulae, or lists of parts may also be written out and subsequently read. The rest of the presentation uses a natural flow of language to develop the ideas in a conversational fashion. The object of extemporaneous speaking is to allow for adaptation.

It may be useful to try several examples for ideas, or to try different combinations of examples, analogies, definitions, etc., each time you practice. Don't just read the outline—try elaborating each of your ideas in a different way. Remember, one of the primary purposes of the practice is to familiarize yourself with the organization, so that you will be less note-dependent during your speech. Try to progressively reduce the amount of notes you use during practice. Effective speakers often have their notes down to a few key words or phrases, plus the supporting quotes, etc., so that a very quick reference to the notes is sufficient to put them back on track should they become lost or sidetracked.

Most persons don't have too much difficulty finding the words to express their ideas in informal situations. However, in informal speaking we often ramble around the topic, use imprecise words, and struggle for specific examples. Preparation of the outline reduces the rambling, allows precision in defining key words, and facilitates the prior development of examples, while maintaining the ability to interact with the audience. You can change the order of the presentation if it is necessary. For instance, a member of the audience may pose a question that relates to material you intend to cover later. It is possible to simply pick up that part of your presentation, answer the question, and return to your prepared pattern. If you see that the audience is not understanding an idea, you can insert another example or analogy.

If you see that you have underestimated the audience's previous knowledge in a specific area, leave out a section of the speech and move on. Manuscripts are very hard to adapt in this fashion, although the use of a half-page format may help. Use extemporaneous speaking to increase the two-way communication. Monitor audience reactions constantly and adjust where necessary.

The use of extemporaneous speeches may help to break down the illusion of separation between you and your audience, especially if your audience views you as somehow different, and detached from their experiences and interests.

# IMPROMPTU SPEAKING

Many times we are called upon to deliver remarks, answer a question, or discuss an issue without advance warning. When we do so we are engaging in impromptu speaking. It is the most frequent and perhaps the most important form of speaking. Many people feel that the ability to think on your feet and speak well is a gift, with which we either are or are not born. This leads to the attitude that there is nothing we can do to improve our ability at impromptu speaking. We believe this attitude is wrong.

One of the techniques effective speakers use is to resist the temptation to immediately spout out responses. In the discussion of the extemporaneous mode we mentioned that one of its advantages was in reducing lack of organization of presentations. Good impromptu speakers take a second or two to contemplate how they will organize their remarks. They may gain this time by restating the question they have been asked, or by introducing their remarks with a summary of what they believe they have been asked to talk about. In effect, these speakers take the time to develop a specific purpose, then inform their audience of their purpose before they begin. Then (hopefully), everyone shares the same frame for the subsequent remarks, and the tasks of speaking and listening are simplified.

As you formulate your purpose, try to think of three main points you will use to accomplish your purpose. This technique is an old debater's trick, and seems to work quite well in many circumstances. Three main points or arguments about a topic are usually sufficient to accomplish a specific purpose in an impromptu speaking setting. You will be amazed at how you can develop the ability to generate these ideas very rapidly if you practice focusing your thinking right away. Ask yourself rapidly, "What is the goal of my remarks? What three things come to mind which relate to that goal?" This concentration of the thinking process is similar to the way the army taught marksmanship—when faced with the need to fire rapidly, trust your basic instincts and coordination. Just be sure to focus everything and point everything toward the target. The first two or three ideas you generate rapidly are likely to be good ones, because they are the ones most salient in your memory.

In advance of your impromptu remarks (as time permits), try the following steps:

1. Structure your remarks with beginnings, middles and conclusions.
2. Clarify your purpose for yourself and for the audience at the outset.
3. Think of the points you want to make and organize them according to their importance or relevance to the purpose.
4. Think of a single thesis statement which introduces each main

point and summarizes it. Start your comments on the point with this statement.
5. Conclude your remarks with a concise summarization of each point you have made, the purpose of the remarks, and seek responses from your listeners that will indicate whether you have presented your ideas clearly and sufficiently.

These strategies can be particularly effective in preparing yourself for a meeting, a session with your boss, or prior to telephone conversations. Even a couple of minutes of planning in advance of any of these communicative contexts will increase your capacity as a communicator.

A particularly important form of impromptu speaking is handling questions from listeners. It is almost inevitable that engineers will be asked questions about their presentations if they allow for these questions. We strongly suggest that you not only allow questions during presentations, but encourage them as a means of fostering better understanding. A brief statement at the outset of a speech, such as, "interrupt me if you are uncertain of anything, or if you have any questions about what I say," may be a good start toward this end. However, this start needs to be supplemented by later encouragement such as, "Is all of this subprocess clear to you? Is there anything I can clarify about this part of the proposal before I go on to the next part?"

When questions are posed by your audience, listen for the whole question. Don't interrupt and don't anticipate the rest of a question, even if you are sure where it is leading. For one thing, interruption is often perceived as insulting. For another, despite where we think questions are going, they often take an unexpected twist at the end. It's not very efficient to save time by anticipating the question, then respond to an erroneous assumption, and finally have to go through a rewording of the question with the subsequent correct response.

There are more reasons for delaying your responses. Questioners often have a way of answering their own questions as they think through the process of asking. Your answer may not be necessary. Even more importantly, if they have thought out their own answer it is usually clearer to them than your response would be. You can merely serve, then, to verify their conclusion. Sometimes you can prod them into developing and stating their own conclusions by suggesting *parts* of a solution, and then asking them to continue.

Waiting until the question is completed also gives you more time to prepare the answer you will give; it allows you to go through the suggested steps of preparing impromptu remarks. We often take the long road to answering questions, fumbling around verbally until our thinking can catch up with us. It is far more effective to be able to immediately respond with the specific answer. If you need time to clarify your thoughts, paraphrase the question back to the questioner. Not only does this give you more time, but it also assures that any other members of the audience can hear the question. Nothing is more annoying to listeners than to play "Jeopardy" with them: "The answer is . . . ", when they don't know or can't surmise the question. Finally, repeating the question allows the questioner to correct you if you have misinterpreted the question in the first place. The advantage of this is, we think, quite evident.

Mentally separate complex questions into their components. Questioners frequently have a way of asking three questions at the same time. We are reminded of a scene from a recent television series, in which Queen Victoria poses a rapid succession of questions to her prime minister, all in one breath. Disraeli responds something like, "To the first question, yes. To the second, decidedly not; and to the third, that is a matter which we shall have to determine." The prime minister kept each sub-question separate in his mind and responded systematically, even though the question was not itself systematic. It may be useful, where possible, to take notes while questions are posed in order to keep track of their complexities.

Admit it when you do not have the answer to questions. Audiences do not generally expect that all of us have total command of all facts and figures on a topic. It is better to simply indicate that you have no answer than to fabricate a response merely to maintain your appearance of competence. Your face-saving efforts may result in your being caught with inaccuracies or contradictions. If you can, assure your audience that you will obtain the desired information and communicate that information to them as soon as feasible. Follow-up is critical.

More and more engineers, as well as other speakers, find themselves making presentations to audiences that are relatively hostile to their proposals. Often this occurs at environmental impact hearings, before legislative bodies, or at open community hearings about proposed technological programs. Many times the hostile audience members pose highly loaded questions. That is, the questions are designed less as a means of gaining clarification from you, and more as a means of gaining an airing of the questioner's own position. There may be no reasonable answer to such questions. Indeed, there may not really be a question to answer.

When faced with these situations, there are several things which we would suggest. Try to determine some part of the question or statement you can clarify, then attempt to move on. Don't merely ignore the question. The rest of the audience won't remember that the questioner didn't pose a reasonable question, only that you were unwilling to respond to someone who opposed you. This hurts your credibility. Try to refocus or reinterpret the question to help you achieve your purpose. Therefore, do not ignore a question you see as irrelevant. It may be more useful to say, "While some of what I hear you asking seems a bit off the point of my remarks, I think that the portion of your statement about sub-proposal X bears clarification." Then restate your treatment of the material in that sub-area.

Recognize that some opponents are prepared on very specific issues.

They may have greater knowledge in that area than you, or greater command of specific figures than you have at your fingertips. While it is often seductive to engage in debate with opponents when you are confident of your position, they may just be setting an ambush for you. Be cautious of debating with those persons who focus on only one particular aspect of the presentation. Try to point out to your audience the broader perspective on the issue—which you are addressing—and move to other questioners if you can. Tell your audience that: "This question addresses only a narrow part of the issue, and, while it is worthy of our attention, we need to discuss other aspects as well." It won't always work, but it may prevent the entrapment these questioners had planned for you. Debate may be useful to your ego, but not to your cause. In many cases, even if you muster effective logic and argument to win your point, the perception of the audience will be that the technologist is the Goliath who did battle with the little David. Even when you win you lose, since the audience sympathy goes to the little guy.

Answer questions that are legitimate quests for knowledge or clarification. Try to restructure unclear or leading questions to match your goals, and avoid debate over minor or specialized aspects of your presentation.

## *DELIVERY*

A final few words on the delivery of any presentation. Audiences like to feel as though you are involved with them, and that the communication process is personally addressed toward them. We cannot overstate the importance of enhancing a sense of involvement with your audience. At the very least, you should make every effort to maximize your eye contact with each member of the audience regardless of your mode of delivery.

No audience will gain much from an oral presentation if they haven't been able to pay attention to it. Use of good, vivid examples and analogies helps the presentation come alive, as do good visual aids. Another important ingredient for maintaining attention is the tone of voice you use in delivering your presentation. No one can tolerate a repetitive vocal pattern for long. Vary your voice—louder, softer, slower, faster, higher, lower—throughout your presentation. Change is critical for attention. Change can also be enhanced through physical movement on your part. We are not saying that you should charge about the room, dance, or jump about on the tables. We are saying that you needn't be a department store mannequin, perpetually frozen in the same position, legs locked stiff, arms glued to the lectern or in pockets, jaw clenched, eyes fixed at a point either directly above or directly below you. *Look at your audience.*

Our last discussion in this chapter concerns something which is universal to some degree in all speakers: anxiety. When we have to speak to an audience all of us tend to feel as though the elevator stopped suddenly on its way up and our hearts (and stomachs) are in our mouths. That anxiety actually can help us, by giving greater dynamics to our voice and physical behavior. It also sharpens our senses, so that we are more aware of the responses of our environment to us. We can use that sensitivity to monitor our audience and adapt our presentations. If we become overly sensitive we begin to feel nothing but negative response. We can overcome this through experience and by correcting our mistakes. In time, we become aware of the fact that the negative responses we perceive are the result of our overly sensitized imagination, not of the audience. Audiences are remarkably unaware of the level of anxiety of speakers. Moreover, they appreciate that most people are nervous in these circumstances, and are very tolerant of many nervous quirks.

With speech anxiety we are much like a motorcyclist riding over the motocross course. Whatever the motorcyclist's eyes focus on, that's where the motorcycle goes. The trick is to look where you want to go, not where you want to avoid. For the speaker it is the same. If you have practiced, or at least know firmly in your mind what your purpose is, then you can focus on that goal—not on all the possibilities for mistakes. Keep your mind busy determining what you are going to do.

Finally, be aware of your breathing. As we get anxious we get tense, and our breathing tends to get shallow. We suspect (without direct scientific verification) that when our breathing is shallow our brains don't get enough oxygen, and we get stupid. When it is your time to speak, pause, exhale the built-up carbon dioxide in your lungs, and take a few regularly paced breaths. It probably won't make you smart, but it may avoid the opposite effect.

## *CONCLUSION*

Oral reporting comprises, either formally or informally, a critical communication skill for engineers. More and more, the engineer or technologist is faced not only with informing others, but with advocating technological proposals. Effective oral communication begins with identification of the audience and of the goal of the presentation. Preparation, no matter how brief, serves to enhance your ability to get your ideas across in an effective manner. In general, a balance between well organized conversational communication and specific facts, figures, authorities, and graphic aids will best facilitate both informing and persuading.

## Questions and Activities

1. Develop a brief (five minute) explanation of a current project you are involved with. Select a specific purpose, develop an outline in which you apply a specific organizational strategy, then develop a set of notes from which you can speak.
2. Deliver your speech to an imaginary audience. Use a tape recorder (video, if available) and review your performance. Did your voice sound natural? Was there a tendency toward repetitious vocal patterns and unnatural pauses?
3. Having memorized the general plan for your brief speech, deliver it again, only this time with no notes to prompt you. Try to simply talk about your material as you would informally. How do the vocal patterns change? Do you sound more natural? How would you improve your delivery, given your two attempts? Try again, using notes only as necessary.
4. Review your speech outline and the recording. Did your speech have an adequate introduction and conclusion? What elements helped you gain and maintain the attention of the audience? Did you use appropriate analogies and descriptions in the body? Did you indicate sources of support to enhance the credibility of your ideas?
5. Think of a speech you have heard lately. Did the speaker do an adequate job? How could the speech have been improved, both in structure and in delivery?

## References

1. Zelko, H. & Dance, F. E. X. *Business and professional speech communication* [2nd ed.]. New York: Holt, Rinehart and Winston, 1978.

# CHAPTER 4

# Written Communication

While most of an engineer's communication time may be spent in the sending and receiving of oral messages, written messages are also critical for professional success. The following chapter will attempt to describe some of the specific techniques of writing which are applicable to practicing engineers. We will not discuss in any detail the very basic skills of writing which are the prerequisites for effective technical writing. The ProCom book *Better Writing for Professionals: A Concise Guide*, Carol Gelderman, and other general writing guides and texts can assist you in developing your general abilities as writers.

In many ways the general principles of technical communication presented in Chapter 2 of this book, and the concepts of effective oral reporting discussed in Chapter 3, apply to written communication as well. Audiences must be analyzed, techniques to clearly describe technical concepts must be applied, and organizational strategies must be employed which enhance the reader's interest and comprehension. There are some specific forms written reports typically take in technical settings, and there are some specific guidelines for writing which we believe you should understand.

Effective written communication requires, as does oral communication, that writers have a clear idea of what they wish to communicate to a specific audience. Without clarification of these communication goals, there is little likelihood that your readers will be able to follow your ideas. Having a clear specific purpose doesn't always guarantee understanding, but lack of purpose almost always assures failure to communicate effectively. Without a purpose in mind, it is likely that your choice of content will be both too broad and too narrow; you will likely include information that isn't needed and forget information that is essential. To use an old analogy, effective writing is like an efficient machine: it contains no unnecessary parts, and all the parts that are needed are present and fit together properly. Just as an engineer cannot design an efficient machine without being aware of its intended function, the unaware writer is likely to devise a written report that resembles a Rube Goldberg contraption. A poorly planned report takes twice as much effort on the part of the reader and the writer as a well planned efficient report.

Without a goal or purpose in mind, which links the ideas of the writer to the actions desired on the part of the readers, there is little hope that the parts of a presentation will work together in an efficient fashion. Organization requires an understanding of the appropriate content as it relates to the specific audience.

Of course, the mere knowledge of the purpose of communicating is only the starting point. Even when we know where we are going, we often get lost along the way. We are going to offer you some suggestions which will assist you much as a road map does. Anyone who has traveled knows that there is often more than one way to get from point A to point B. Each route offers its specific advantages. One may be more scenic, another may be faster, a third may simply be safer. While technical writing seeks efficiency and hopes to achieve results with predictable assurance (safety), it has typically avoided the scenic route by maintaining an impersonal and dry style. We contend that the destination can be reached through combinations of safety, speed and scenery, and that one often leads to another in surprising ways. Just as the quickest route from Seattle to Tokyo isn't straight westward, sometimes an element of scenic writing can do a lot to shorten the route to understanding. Once again, this is especially true of the use of analogies and effective graphic devices. In writing, as in engineering, the desire for efficiency can only go as far as the successful accomplishment of the desired function allows. It does little good to pare down our writing for the sake of efficiency only to discover that the purposes of the writing are no longer accomplished.

## GENERAL SUGGESTIONS FOR WRITING EFFECTIVENESS

The techniques for selecting and organizing material presented in the previous chapter are even more important when we write. Identify your purpose, gather information, systematize that information, and select from it that which suits your purpose. List that material and evaluate each idea. Discard those ideas which are only slightly related to the purpose, and attempt to determine if the information which you have gathered is sufficient. If it isn't, then take the necessary steps to acquire the needed ideas; review the resources in the area (or conduct your own tests) that you can base your conclusions upon. Organize your ideas according to the needs of your readers and according to a logical systematic pattern. Finally, outline your material and determine how you can tie the various parts of your outline together. As we have suggested before, look to see whether you have included an introduction that tells your readers what to expect, and conclusions and summaries that tell them what you have told them.

As you begin to write from your well-developed outline, we suggest that you keep three things in mind:

1. *Good writing almost inevitably means good rewriting.* No one gets their ideas down onto paper perfectly the first time around. The best writers recognize this, and go over and over their writing to make it better. We don't mean just reviewing the writing for typographical errors or misspellings (all of which is essential). We mean a serious effort to determine whether the ideas have been presented as effectively and concisely as possible. None of us can escape the necessity of editing and rewriting. Just as engineers must constantly check and verify the results of their tests, so must writers check their writing.

    While it obviously takes more time to rewrite, the savings in increased efficiency through greater shared understanding make the effort worthwhile. As important, perhaps, is that this effort on your part helps to foster positive perceptions of *you.* Writing that is hurried and unedited leads readers to believe that you are careless and unconcerned in other professional activities.

    Finally, one of the greatest problems many writers have is getting their ideas out of their heads and onto paper. If you are overly concerned with expressing your ideas perfectly the first time around, it is likely that you will lose the flow of ideas; you'll forget important points or particularly good examples as you struggle to perfect the mechanics of your writing. Working from an outline helps to avoid this difficulty, but can't totally overcome it. You are writing primarily to communicate your ideas. Get those ideas down, and then rework them. Even short memos will be improved if you rewrite them.

2. *If possible, read your material to yourself out loud.* We recognize that it isn't always feasible to read everything you write out loud. Friends and colleagues may begin to wonder about your talking to yourself. We do think that you can read

much of your material very quietly, though. When we write, it often comes out like our thinking; that is, we mentally fill in some blanks for ourselves. Because we know what we are thinking about as we write, we don't consider how others will know our thoughts. When we actually *hear* the sound of our sentences, we often detect that words or ideas have been left out which help others understand us.

When we read our material silently we often fill in words that ought to be there but aren't, or ignore words that are there but shouldn't be. Simply put, we tend to be more careful readers when we speak things out loud.

Perhaps most important of all, despite many years of studying English and writing, most of us don't consciously know all of the rules of correct grammar and usage. Nonetheless, most of us are able to detect usage which doesn't sound right. When we hear language used incorrectly, we may not know why it is wrong, but we are able to work with it until it sounds better to us. Again, this won't always prevent poor writing, but it is one technique that often helps us and others who have tried it. As often as not, when we ask students to read aloud poor sentences they have written, they are immediately able to detect the problem and correct it.

3. *Always write with a specific reader in mind.* One of the great disadvantages of written messages is that we are often unable to receive immediate feedback from the readers. In fact, we often aren't sure exactly who the reader is likely to be. Or there may be many potential readers, all of whom aren't there to seek clarification or amplification from us. We have emphasized throughout this book that one of the keys to successful communication is the ability to adapt communication to the needs and characteristics of our specific receivers. When the receivers aren't present to provide feedback, and when we are unsure in our minds of these receivers, we seem to slip into writing patterns which are clear to ourselves, but not to anyone else.

Many writers compensate for this problem by selecting a single person who is representative of the potential readers, and then adapt their writing to suit this person. Concentrate on how this one individual would be likely to understand your ideas, what this person's needs are, and then write imagining the likely responses from the target individual.

If you are writing for the general public, try imagining how you would explain your points to your neighbor—the one who can't mow his lawn because he can never get the power mower to work right (and when you lend him yours it comes back broken). What would it take to get your point across to him, even if he were there to ask questions? Where would he be likely to ask questions, given what you have written? Strengthen those areas of writing you anticipate this individual having the greatest difficulty understanding.

If you are writing for fellow technologists, picture just one of these persons, and write for that person. Try to select a target who typifies the audience, or who represents the type of reader you are most concerned with reaching.

We've given just a couple of examples, but we are sure that you can think of many other cases where you could select just one person to direct your efforts toward, which in turn would help you reach others more effectively. Beware, however, because this technique can lead to difficulties if you are writing materials for several readers who have very different expertise and orientations. It may sometimes be worthwhile to write for one individual the first time through, then reread and edit your writing with other individuals in mind, to check whether you have met their needs as well. Try to put yourself in their place, and critically evaluate your own efforts to communicate.

One final caution before we move on: It is sometimes necessary when writing technical material to mentally picture a target who is *not typical* of the general reader. A specific instance of this is when you write technical instruction manuals. Here your task is to write for the *least capable potential user* of the product or process. While the more capable users may be bored with this level of instruction, it is far easier for them to move through the unnecessary material than for others to try to figure out things on their own. The risk of idle or damaged equipment, or a permanently destroyed environment, may be too great to allow for assumptions of competence on the part of your receivers. It's usually wiser to err on the side of greater detail than less; but the ultimate decision of balance must come about as a result of your careful consideration of the audience and the risks. What seems simple and obvious to you rarely makes as much sense to anyone else.

## WRITTEN STYLE

Writing differs from oral communication in several important ways. Developing a sense of these differences can help you not only to write better, but also to determine whether the message you wish to communicate is best transmitted in oral or written form.

We stated earlier that written communication doesn't allow immediate feedback. You don't have any way of knowing whether your explanations or descriptions have succeeded in accomplishing your purpose. As a result, you can't easily correct misunderstandings, or add clarifying material later.

To overcome this, you must build into your writing a somewhat greater depth of description and explanation. Use more examples, analogies, graphics and definitions when you write. (Don't get carried away and beat a dead horse, however.) The material you hold in reserve as a "fall back" (if your initial efforts to communicate fail) in an extemporaneous speech often needs to be included in written presentations. In the absence of immediate audience responses, this insures your success.

When you are providing this secondary support cue your reader by prefacing it with statements such as, "A second example of this is . . . ," or "for further clarification." This allows the reader who has grasped your ideas to skip ahead.

Readers don't always read everything in reports or letters. You can help the reader who is scanning your material by providing headings for sections of your report, and by including clear introductions and summarizations to the various parts of the written material.

Each part of the report should begin with a brief statement of the purpose of the section and a preview of the material which will be included. The reader can immediately determine whether the following section contains the information they seek. An effective, concise introduction to a section also helps to assure that you know what you are attempting to accomplish as well.

Written communication usually allows for the presentation of more complex material than oral communication. Since readers may proceed through written material at their own pace, they can slow down for complicated material, reread sections, and refer back to earlier sections to check for background information where needed. Lists of data or statistical and procedural descriptions—which would be incomprehensible because of limited oral memory—may be far easier to comprehend when written, especially through the use of graphics such as summary tables. Graphics may be more detailed in the written form, since their legibility is less of a problem than when projected for a large audience. Finally, the written mode allows for a permanent record of the material to be preserved in a relatively accessible and efficient manner. Larger volumes of information may usually be processed in a fixed time in the written mode than in the relatively slow oral delivery.

# TECHNICAL STYLE

Technical material is usually written in a fairly specific style. This style is summarized by four descriptive characteristics: *Impersonal, Objective, Concrete,* and *Free of Jargon.*

## *Impersonal and Objective Style*

The first two, Impersonal and Objective style, tend to go hand-in-hand with one another. The idea of impersonal writing is not an absolute requirement of technical writing, but many times it serves to enhance the goal of objectivity. Usually, readers of technical material are uninterested in subjective, personal opinions about the technical topic. They are concerned with verifiable, systematic observations and conclusions. The inclusion of a personalized style, or the insertion of personal pronouns (I, We), may decrease the perception of objectivity. For this reason, traditionally, technical writing has avoided such usage. Of course, the impersonal style often becomes dry and uninteresting to the reader. Obviously, a balance must be found. Writing that reads like an autobiography of the researcher or technologist, rather than a description of a technical process, may lead to perceptions of the author as self-serving. A sense of false objectivity, on the other hand, with no mention of the roles of the individuals making the decisions and conducting the operations, may lead to the sense that there is no involvement or commitment on the part of the writers and researchers.

Regardless of the degree to which the writer is personal or impersonal in style, the need for objectivity remains. The inclusion of information which describes the procedures and results of your investigation in detail is essential. Assertions and conclusions must be based upon the data from the study you have conducted, or upon previous research and theory. The basis for the conclusions must be evident to the reader, and must be well-supported by citations of either the relevant data or previous research, so that the reader may verify your conclusions. When you must "call a close one" and give *opinions,* be sure that the reader is able to distinguish your informed opinion from factual statements. If this distinction becomes unclear, the reader may soon lose faith in all of your material.

We have already spent some time discussing the need for conciseness in writing. Earlier, we pointed out that an analogy may be drawn between effective writing and efficient machines—neither have unnecessary parts. Unneeded words lead to wasted time. Get to the point, state the point, and move on. Readers of technical material are rarely impressed with elaborate prose that exhibits a wide vocabulary, but a shallow understanding of the essentials. Review your writing. If a word isn't needed—remove it. A sentence—remove it. A section—remove it. Keep asking: Does this help accomplish my purpose?

## Concreteness

Concreteness refers to the ability to be specific in describing or explaining your material. The effective technical writer avoids abstract descriptions. Compare the statement, "This car gets excellent gas mileage," to the statement, "This car is in the top 10 percent of all tested cars according to E.P.A. testing procedures." Words such as *excellent, good* or *poor* are inevitably abstract; they are meaningful only in relationship to specific standards. The words are only useful, if at all, when the standards are explicit, and when the standards are themselves valid. A statement that a car gets good gas mileage is obviously subject not only to the testing procedures which lead to the conclusions, but also to the standards by which the procedures are evaluated. We used to think that the 25 miles to the gallon our 1973 pickup gets were "good" figures. By today's standards, the same figures aren't as good as they once were.

There are many other examples of such abstract terms. Among them are: *high, low, majority, minority, several, occasionally,* and *frequently*. The list could go on and on. None of these words is inherently a "bad" word. There are times when they reflect a reasonable description of events or concepts. Nonetheless, where possible they should be scrupulously avoided by the technical writer.

## Free of Jargon

Our final technical style criterion is that your writing should be free of jargon. Every profession develops its own specialized vocabulary. This is often an efficient "shortcut" for communicating concepts that would be time-consuming to elaborate in everyday conversation. This jargon varies in its ability to be understood by others. While the term *scrap* in a foundry has an intuitive meaning even to outsiders, terms such as *soup, crash, dump, pot, core,* or *ram* may have very different meanings to others. The use of jargon requires very careful audience analysis. We are often unaware of the jargon in our vocabulary, however, and equally unaware that others don't know the meanings of words which are so much a part of our everyday vocabulary. You must be sure that you have defined terms adequately for your audience, and that you are not assuming understanding inappropriately. In our experience, we have found that there are often sharp distinctions in the meanings of terms, even within a given field. When you write for a general audience, use the precise terms you need, rather than the shorthand jargon. This is, of course, especially important when you communicate with nonspecialists. Remember, we are all becoming more and more specialized in different areas, so that even the basic critical concepts and terms of many fields are totally unfamiliar to specialists in other areas.

Two special forms of jargon which have become increasingly prevalent are the use of acronyms and initializations. Acronyms are words which are derived from the first letters of several other words. Initializations are just the first letters of other words, but aren't themselves pronounceable words. One of our colleagues related the story of a friend told her about a planned trip. Our colleague was deeply envious of her friend, whom she envisioned basking on tropical beaches sipping rum concoctions, only to discover that her friend wasn't in Nassau, but was at NASA in Houston—a far cry from a tropical vacationland.

This is an extreme example of the misunderstandings which may result from acronyms, but it does indicate that this type of jargon can confuse readers or listeners. A colleague, as a student learning to use a computer, was frustrated for hours trying to accomplish some file handling which always resulted in a FURPUR statement from the computer. It sounded so much like a drastic mistake that he kept checking to determine what was wrong, and kept resubmitting the run. Finally, he learned that the FURPUR was a perfectly normal statement standing for File Utility Routing/Program Utility Routing and was exactly the message that should result from the requested operation.

An example of misuse of an initialization came across one of our desks recently. A memo requested that recipients nominate candidates for office in the CSAW. This type of request is fairly common, and would normally have been disregarded unless the CSAW was particularly important to us. Unfortunately, to this day we have no idea of whether the CSAW is important or not. The person who wrote the memo neglected to inform us of what CSAW stands for. Once again, failing to realize that what was perfectly obvious to him was not so obvious to others led to a lack of effective communication.

Such confusion is inevitable if we aren't scrupulous in describing the underlying words from which acronyms and initializations are derived. Be sure that you indicate this information the first time that you introduce the acronyms you use. Never assume that even the most experienced readers in your area will interpret that acronym correctly on their own.

A special problem of style in technical writing, which falls outside of jargon usage, but is closely related, is what a former English teacher of ours used to call "highfalutin, overpollutin' words." By this he meant words which, while not incorrect, could easily be replaced by simpler, more concrete words. In the technical world, certain words and phrases have become almost jargon by their constant use. Just a couple of examples will, we think, give you an idea of what we are cautioning against: *Activate* or *initiate* instead of *begin; terminate* instead of *end; consensus* instead of *agreement* (unless you really mean consensus); *in the event of* instead of *if; commensurate with* instead of *fits* or *agrees with;* and on and on the list could go. The educated reader may know all of these words, and a certain amount of variety of word use is desirable, but in general the use of these words or phrases only makes the reader work harder to gain simple understanding. We believe in a simple maxim: Decrease reader effort—increase reader understanding. (Note that we didn't say *minimize* and *maximize,* two of the most over-worked of the highfalutin' words.)

Our description of a technical style is offered as a set of suggestions that address what we feel to be some of the more critical problems in writing. As with all forms of communication, concentrating on making your reading understandable for your specific audience will do much to enhance your style. Now, we turn our attention to specific content elements for the writing of technical reports.

## PARTS OF A FORMAL TECHNICAL REPORT

No single format is likely to be applicable for all technical reports. Different components are necessary, depending on the problem area and the purposes of the report. Once again, the needs of the reader play an essential part in determining what information goes into any report, and how that information should be organized. Feasibility reports may take the form of the problem-solution pattern discussed in the chapter on oral reporting. Progress reports may be organized in a different fashion, such as a chronological pattern, or according to the different aspects of the project being described.

Because of these many possible configurations for reports, this section provides a description of several of the key elements common to most reports. Some reports may not include all of these, and others may require unique elements not mentioned in our limited space. A typical report might contain the following:

- A. Letter of Transmittal
- B. Title Page
- C. Abstract
- D. Table of Contents and Lists of Illustrations
- E. Introduction
- F. Body
- G. Conclusions and Recommendations
- H. Bibliography
- I. Appendices

In the following sections we will give some brief descriptions of each of these elements.

## A. Letters of Transmittal

A letter transmitting the report simply indicates for whom the report is intended, and by whom the report was prepared. In addition, the letter usually indicates why the report has been sent; for instance, whether the report fulfills a specific contract, or was requested by the addressee. In these cases, the appropriate grant or contract number would be indicated, or the date of the specific request for the report.

The letter may also include a brief section pointing out any critical sections of the report that are particularly relevant for the intended recipient. You may also indicate any action you expect or require on the part of the recipient. When reports are transmitted within an organization, the letter of transmittal is replaced by a memo containing essentially the same information. Included below is a sample of a letter of transmittal for a feasibility report.

### SAMPLE LETTER

3525 N. Mill Rd.
Glendale, WI 53207
December 9, 1982

Dr. Richard H. Arthur
Director, Communication for Engineers
Department of Communication
University of Wisconsin-Milwaukee
Milwaukee, WI 53201

Dear Dr. Arthur:

This report on the sewage treatment process at Jones Island is submitted to you in compliance with your assignment for Communication for Engineers. Details of this assignment were specified in the course syllabus and in your handout on November 7, 1982.

The report defines sewage characteristics in the Metropolitan Sewerage District. Next, sewage flow figures and a description of the system of sewers, etc., are provided. A discussion of the methods of treatment and of handling excess flow follows. Problems with current methods are identified and analyzed.

Several alternative methods for handling sewage are presented and evaluated against criteria established by the group after consultation with engineers from the Sewerage District. Our recommendation for the construction of deep tunnel storage facilities is presented for your consideration.

Sincerely,

Carl Porter, Secretary
Group Four

As with all business letters, the letter of transmittal should use an appropriate form. Note how, in the example, the margins are balanced both left and right and top and bottom. The example we have provided uses the *full block style,* in which all of the elements of the letter are begun on the left hand margin. Another commonly used form places the date and the closing toward the right hand side to balance the appearance of the letter. Either style is widely accepted in the business world. The aim of both styles is to create a letter which has a neat and pleasing appearance to enhance the reader's perception of the quality of the ideas presented. A sloppy-looking letter leads the reader to expect sloppy work, and may color the interpretation and acceptance of your ideas and suggestions.

Be particularly careful to proofread your letters, whether they are part of a report or for any other business purposes. In our experience, we have heard more complaints about misspelled words and outright sloppiness in business letters than any other single complaint from businessmen. We think that perhaps this problem is no more important than many other chronic communication difficulties, but recognize that the consistency with which we hear this complaint suggests how important others consider this issue to be. Since it is such an important issue, it is well worth your effort to spend some extra time to make sure that this first part of a report is perfect. Even if prepared by a secretary or another professional, check it over yourself for any errors. Take out your dictionary if any spellings are in doubt. Leave nothing to chance.

As a help to you in your letter writing, whatever the purpose, we include the following guidelines.

## Guidelines for Writing Letters

### The Parts of Letters:

In general, all business letters contain several critical parts. Each of these parts is designed to provide critical information or to perform a specific function. The main parts of a business letter are:

1. Information about the sender (heading)
2. Inside address
3. Salutation (Sometimes optional)
4. Body of the letter
5. Closing
6. Signature

**Heading:** In the heading of the letter, you provide the recipient with information about yourself. Specifically, the heading should contain, at a minimum, the sender's street address, city, zip code and the date.

*Example:* 4640 N. Woodruff Avenue
Whitefish Bay, WI 53211
September 2, 1982

When the letter is typed on company letterhead, it is only necessary to add the date. When letterhead is not used, it may be desirable to include the name of your company as the first line of the heading.

**Inside Address:** The inside address contains the name and title (if appropriate) of the person or organization to whom you are sending the letter. This is the formal name, not a nickname. Following this, the address is written.

*Example:* Dr. Robert Payton, Chairman
Wolfshead Design, Inc.
3552 N. 52nd Street
Milwaukee, WI 53207

**Salutation:** The salutation, or greeting to the recipient, should contain the name of the recipient, if it is at all possible. A little extra time to determine the appropriate addressee, by name, personalizes the letter, increases its impact, and shows your attention to detail. If you are a close friend of the recipient, it is often considered acceptable to use a nickname.

*Examples:* Dear Mr. Culvert or Dear Carl

If you are unable to determine the actual person to whom you are sending the letter (as when you are writing to a company) several alternatives are available. Traditional form demanded a *Dear Sir:* salutation. However, with growing recognition of changing gender roles, the traditional form has become unacceptable to many persons. An alternative is *Dear Sir/Madam:*, but even this has an element of awkwardness. Currently, many persons in business choose to delete the salutation entirely if the gender of the recipient is unknown.

*Examples:* When the gender of the recipient is known, but not the name; *Dear Sir:* or *Dear Madam: Do not* use *Dear Sir* when the gender is unknown.

**Body of the Letter:** The message you wish to convey is the body of the letter. The opening paragraph should contain a concise statement of the purpose of the letter and the action you desire from the recipient. The remainder of the letter's body contains elaboration and explanation of that specific purpose which you described in the opening. The final paragraph should contain a summarization of the letter's content including a restatement of the purpose of the letter and what you expect the recipient to do. There should be only one main idea expressed in each paragraph of the letter. Keep paragraphs (and sentences) short and try to hold the letter to only one page.

**Closing:** The closing formally signals the end of the letter. It conveys respect for the recipient.

*Examples:* Yours, Yours Truly, Sincerely, Respectfully,

**Signature:** Below the closing, space is left for you to write your signature, then the following information is written: Your typed full name (and any degrees, or professional certifications you use) and your title in your organization. *Example:* Yours,

    John Trestle, PH.D. CPA
    Comptroller

### Letter Parts Which May Be Needed:

**Stenographic Reference Line:** When the letter is typed by someone other than the author, it is customary to indicate this by the inclusion of a stenographic reference line. Two sets of initials are included in this line: the initials of the author of the letter, in upper-case letters, followed by the initials of the typist, in lower-case letters.

*Example:* RHA:bk or RHA/bk

**Notation of Enclosure(s):** When additional information or materials are to be included with the letter, a notation to this effect is placed at the bottom of the letter. This signals the recipient that there is additional material and prevents discarding envelopes which may contain items which were not initially removed.

*Example:*   RHA:bk     RHA:bk
              Enclosure   Enc.

When you are enclosing materials it is usually wise to indicate this somewhere in the body of the letter as well as with the enclosure notation.

### Letter Format:

As with any presentation you make, the form and delivery of letters conveys to the receiver important impressions about you. Sloppy letters convey an image of a sloppy indifferent sender. All letters which go out over your name should be carefully proofread and checked for proper form, *by you*. Don't trust others to safeguard your image.

Several formats are considered appropriate for business letters. We will describe only one of these formats: the full block form.

**Margins:** Margins should be equal, left to right, and top to bottom. Margin on the left is usually one inch, but this may be adjusted for shorter letters so that more of the page will be filled; i.e. increase the margin for very short letters. The right margin should be approximately equal to the left and the ends of lines should be even with one another, if possible. Word processing facilitates this right-hand justification, but if processing is not available, make every attempt to keep the right margin even.

Try to keep the letter in the middle of the page. If the letter is short, start it lower on the page. Top and bottom margins, excluding letterhead, should be equal.

**Spacing:** The date is either a part of the heading or is placed where the heading would go to achieve a top and bottom balance for the letter. Likewise, if you are using a full heading it is placed to allow an equal margin above it and below the signature.

The inside address is placed four spaces below the heading or date.

The salutation, if included, is placed two lines below the inside address, and the body of the letter starts two lines below the salutation (or the inside address, if no salutation is used).

In full block form, the beginnings of paragraphs are *not indented.* Each paragraph begins at the left hand margin, is single spaced and is separated from the next paragraph by a double space. If the letter is very short and contains only one paragraph, it may be double spaced.

The closing is placed two spaces below the last line of the body and the handwritten signature is directly below the closing. In full block form, the closing is begun on the left margin. Four to six spaces should be allowed between the closing and the typed signature. Two spaces separate a stenographic notation from the signature line (or title if used).

## B. Title Page

The title page repeats some of the information stated in your letter of transmittal: title of the report, who prepared it, who will receive it, and relevant contract information. All of this is necessary, since the letter of transmittal may be separated from the report prior to distribution within the recipient's organization.

Center the material on the page, and spread the information out to balance the *white space* across the page. Take particular care in selecting your title (if it has not been specified by your contract or agreement). Be sure the title clearly and accurately reflects the specific purpose of your report. Titles are one of the means by which potential consumers of your report identify it. Try to be sure that your title contains *key words,* which identify the report as belonging to the related body of information in the area under investigation. These key words are principal tools for investigators seeking information through computer searches of bibliographic sources.

## C. Abstract

The abstract is one of the most important parts of any report. For one thing, it often determines whether your report will be read. If readers of the abstract are not interested by the abstract, or cannot gain a clear sense of the contents of the whole report, it is unlikely that they will go any further.

*Figure 4-1* **Letter formats**

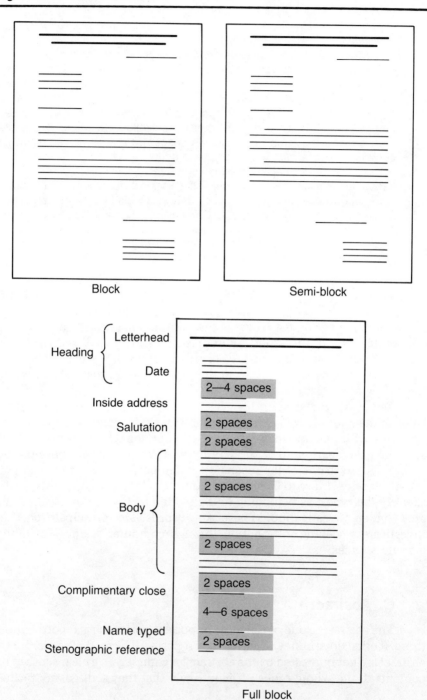

Readers expect a clear description of the report, and, where appropriate, brief descriptions of the problem investigated (and how it relates to other studies), the methodology used, general results, and the specific conclusions and recommendations offered.

All of this information should be in the most concise form possible. This is not the place for a complete explanation or justification of your procedures and conclusions. As briefly as possible (while assuring that the reader will have an accurate picture of your efforts), tell the reader what you did, why you did it, how you did it, what you found out from it, and what you think should be done with it now.

Since modern information retrieval systems often rely on the contents of the abstract, in addition to the title of the report, the abstract is doubly important. Retrieval systems search for key words in the abstract, so there must be reference to the commonly used terms for the specific problem area under study. For this reason, most abstracts usually contain a separate listing of the key terms from the body of the abstract and from the report. If the terms that will help a future researcher find your material aren't present, then your efforts may be lost in a matter of very little time.

Look at the reports of researchers that served as sources for your own investigation. Identify the terms they used, and be sure that your list of terms cross-references at least some of the relevant words.

A good way to practice abstract writing is to develop a *precis* of a report. A precis is developed by going through and summarizing the contents of each unit of the report. It may, if it is very detailed, extract the main thought (or thesis sentence) of each paragraph without the accompanying elaboration. A less detailed precis might summarize each subsection of the report. From this precis you select those statements which are most essential to the understanding of the report. This process helps to assure that nothing critical has been overlooked in your abstract. It also assures that the abstract contains accurate statements drawn from the report itself, rather than from your own subjective memory of the material. The ideal technical report abstract should contain *no more than 250 words*. Experienced abstract writers (and readers) suggest that after your initial draft of an abstract, you should go back and trim it by 30 percent to 50 percent. They contend that there is always "fat" in an abstract that can be cut away. Four sources may be particularly useful to you in preparing abstracts according to industry standards:

1. *The Guide to Source Indexing and Abstracting of Engineering Literature*, published by the Engineer's Joint Council.
2. *Thesaurus of Engineering Terms*. A guide to terminology which can assist you in using correct keywords for easy reference and retrieval.
3. *Guidelines for Cataloging and Abstracting*, published by the Defense Documentation Center.
4. *Directions for Abstractors*, published by the Chemical Abstracts Service.

## D. Table of Contents and Lists of Illustrations

These are somewhat self-explanatory. The table of contents should indicate the page number of each major section of the report. If you have used subheadings within each section (as we recommend), indicate the page number of these. The object of these tables and lists is to assist the readers in locating the information they seek. Be sure to include the location of this key information.

## E. Introduction

The introduction to a written report serves the same functions as the introduction to an oral report. An effective beginning usually includes description of the purpose of the report, the scope of the investigation, and any relevant background material (to help the reader gain a sense of the place the report holds in the larger area it addresses). The methodology used to conduct the study is usually explained briefly. Key terms are listed and defined to establish a common understanding with the reader at the outset.

In many cases, a brief summary of the conclusions and/or recommendation is included; especially if the report is submitted to laymen, who may be relatively uninterested in the detailed description of how you arrived at the conclusions, and are seeking only *the bottom line*—what this report asks them to do.

So, with any introduction it is useful to stress the importance of the report to the readers and to the general area of research. It is also helpful to include a preview of the remainder of the report. The readers then have a clear idea of which sections are most important to their interests.

## F. Body

The major portion of the report is the body. The body is organized according to the type of study and the needs of the reader; for instance, chronologically to explain a process. The body includes the data generated by your investigation (if you have run tests). Obviously, there seem to be as many strategies for selecting and organizing content as there are different reports. We cannot provide a description of all the possibilities. The important thing is to plan—prior to writing—a logical body, including the material relevant to your purpose and organized to assist your readers. Ask yourself what your readers need to know, and what you need to include to help them know it.

## G. Conclusions and Recommendations

Here is where you draw all of the threads of your report together again. You summarize the material you have presented and restate the purpose of the investigation. You may have already stated your conclusions in the introduction, but repetition of these at the end is important. Many readers automatically go to the end of a report first to determine the results.

The conclusion also includes, as appropriate, the recommendations derived from your reports. The specific things you believe your readers should do must be clearly stated, preferably in a simple step-by-step fashion. Readers (especially managers) expect concrete, definitive, objective statements based upon your research and development. Don't go off on tangents and introduce new data or new areas of concern in the conclusion, except as recommendations for further research. The support for your conclusions and recommendations belongs in the body of the report. A conclusion cluttered with new information or reanalysis of previous material loses impact, and leaves readers confused and uncertain of your proposals.

## H. Bibliography and Footnotes

It is your obligation as a researcher and writer to indicate the sources of your information. When you use someone else's idea (not just direct quotations) you are borrowing their "product," and they must be compensated through recognition of their contribution. It is much like renting a tool. If you don't pay the rental, you have stolen the item. The idea is always just "on loan," and remains the property of the original source.

Indicating the source of ideas and information allows the interested reader to seek further clarification from the original source, or to check your interpretation of the conclusions others have drawn. It may also assist the legal department of your organization in tracing the origin of your proposals to avoid patent or copyright infringements.

There are many forms which footnotes and references may take. The format is sometimes determined by the outlet you seek for your report. Journals have specified guidelines, as do some grant agencies. Consult the guidelines usually provided by the requesting agency or the journal to which you submit your report. In the absence of any guidelines, it may be best to look at the style of footnoting and references used in your organization, and use these as a model.

As a minimum, references contain an indication of the author(s), the title, journal title (if appropriate), and publication information such as date and company of publication, volume number and page numbers. Consistency is the key in documenting sources. Pick a style and stick with it throughout the report. If you write reports a wise investment is *The Chicago*

*Manual of Style*, 13th Ed. (Chicago: University of Chicago Press, 1982) or *Style Manual*, University Ed. (Washington D.C.: United States Government Printing Office, 1967).

## I. Appendices

Technical reports often include appendices. The choice of whether material, data, charts, tables, etc., goes in the body or is collected at the end is, once again, determined by the relevance of the material for the reader and to your purpose. Appendices include *supplementary* material which is not essential for the understanding of your purpose. It is placed at the end for the curious reader, who may wish to check the raw data which you have analyzed and summarized in the body, or examine tables containing less important background material.

Readers are generally referred to the appendices by a parenthetic statement in the body of the text, as in the following examples: "Data were gathered on the load factors (Appendix E) and analyzed," or "Relationships between all of these factors were previously determined (See Appendix B for a summary of all relationships)."

Not all of the components are necessary in every formal technical report. Certainly, any report will have an introduction, a body (with a variety of possible organization and content), and conclusion. Inclusion of other parts depends on the needs of readers, and, to some extent, on the degree of formality of the report. Not all reports are full-blown formal reports, such as we've discussed here. Many reports are quite short and informal in their structure, yet vital communication tools for the engineer. One particularly critical type of informal report is the memorandum.

## MEMOS

Memorandums, or memos, are the mainstay of written organizational communication. It is through memos that we typically inform others of the day-to-day operations and problems in technological settings. Each memo serves as an informal report, in that it doesn't include all of the elements of larger reports, and often is personalized for receivers. However, it contains critical information that aids formal functions within the organization.

In our experience, we have heard more complaints about the quality of memos than any single form of communication. We believe that much of the problem lies in writers' failure to recognize the formal function of these informally written documents. The solution, quite simply, lies in recognition of the importance of memos, both to the organization and to you personally. Memos may be informal, but they should never be haphazard or careless. Each memo reflects your capacity for clear thinking and attention to detail, as well as clear expression of your thoughts.

Among the typical formal functions of memos there are four we believe to be most prevalent. First, memos document action taken. They tell others, *for the record*, what we have accomplished or failed to accomplish. As such, they are vital in informing the other members of the organization of the current status of projects and activities. Managers simply cannot make effective decisions regarding resource allocation, future planning, etc., unless they are accurately informed of where the organization stands at a given time.

Second, memos inform the organization of actions to be taken. Other members of the organization often need to be informed of your plans, so that they may coordinate their planned activities with yours. They also may be able to correct any errors in your planning, if they know about it in advance.

One of the great problems in industry is a tendency to "reinvent the wheel." When others are informed of your plans they may be able to inform you of previous work which has already solved your problem, and provide other helpful supporting information. The National Aeronautics and Space Administration (NASA) used memos for this function particularly effectively.[1] Each division provided memos on their activities. These memos were distributed to all other divisions, which then could provide any information they had related to the proposed actions. If they already had a solution, they said so. If they knew that the proposed action wouldn't work, they were obligated to inform the originator of the memo.

The third function of memos is to seek information directly, to request action on the part of others in the organization, or to provide responses to these requests. Memos of this type may ask questions of specific persons, or are sent as general requests to anyone who can provide the needed assistance. When sent out in general distribution, they allow for discovery of hidden information, which you might not discover through personal contacts with your usual associates. When sent to specific individuals, they typically require some formal response, even if the recipient cannot accomplish the desired action, or provide the needed information. Because these memos virtually mandate responses, they should be used carefully. Don't send them unless you are relatively sure the recipient can give you what you need. Avoid swamping others with an overload of paperwork, which in turn calls for paperwork in response. Memos are costly in preparation, distribution, and response time.

The final function of memos is to clarify and affirm agreements, discussions, or observations. Typically, memos are used as followups to oral communication. When we converse on the telephone with others we have no record of our mutual decisions. The memo serves as a record by summarizing the conversations we have had, and assuring that both parties understand one another as to action to be taken, etc. Many times, memos are sent with a request that an initialed copy be returned to assure the receipt of the memo and agreement on the content. In this case, to some degree, the memo serves as a contract between correspondents.

Again, this usage of memos is subject to abuse. In many organizations, memos of this type abound, with everyone attempting to document each conversation so that responsibility for decisions may be placed on other shoulders. In the military (and elsewhere) this behavior is termed C.Y.A., for "Cover Your Ass." Every conversation is documented, and the flurry of paperwork consumes the better part of some individuals' work time. Record only those conversations which are of particular importance or sensitivity, or where there may be a real need for additional clarification, because of a misunderstanding or the need to note new information.

Before a memo is written, careful consideration should be given as to its real necessity. Send only memos which have a clear specific purpose in the organization. This implies that you must be certain of your purpose, which is the starting point for preparing memos.

## PREPARING MEMOS

As with all the other forms of communication we have discussed in this book, memos require two starting points: consideration of the purpose and consideration of the receiver. Memos should begin with a clear statement of their purpose and of the desired action on the part of the receiver. They follow this information with the essential background for the particular readers. Consider the following memo:

> TO:
> FROM:
> RE: OUR CONVERSATION 11/10/80
> I was thinking about our conversation last week. It was good to hear about your progress. Keep it up. It seems to me you could solve one of the problems by some reinforcing at the central point, as well as other ideas. Tell me how it works. Jane sends her regards to you and Carol. We need to get together soon. Have yours call mine.

This memo exhibits a multitude of problems. We know from the "Re:" statement that this memo will have something to do with a previous conversation, but we know little else. This memo may have made sense to the writer, but we are certain the recipient was confused. The specific purpose is unclear as are many other parts. It isn't certain whether this is a recommendation, a clarification, or a request for action. Will the recipient understand what problem the writer is referring to, and where the critical point is? Does the writer want to get together socially with the wives, or to formally discuss the suggestions?

Compare the earlier memo to this one:

TO:
FROM:
RE: OUR CONVERSATION 11/10/80
Let's set a time this week to discuss an idea I have about your stress problem. How about 10:30 tomorrow? Call my secretary and confirm or get another time.

It seems to me that some reinforcement at Point AD in your plans will be helpful. Give it a look and we'll decide.

Incidentally, Jane says "hi" to you and Carol.

The actual information is right at the top. The recipient knows what he is expected to do, and the memo is clear about the suggestions. Ambiguous memos lead to errors. Be certain yours have a clearly stated purpose.

As in all writing, be concrete and concise. Use words your reader will be certain to understand. If there is any doubt, use another word or define your terms. Especially for memos that serve record-keeping functions, be sure that they are sufficiently detailed to be understood later, when memories are fuzzy. Consider the following:

TO:
FROM:
RE: MODIFICATION TO PRODUCTION EQUIPMENT
You have my approval for the first idea, but let's hold on the other two; I'll take responsibility with M. and inform him. File this as a record of my decision.

How long will it take before no one is certain which idea was first, and which were second or third? While it is common to simplify statements in memos to a telegraphic shortened style (e.g., "Send Q report fast. Need for 10:00 A.M. meeting."), be sure that simplification doesn't omit critical information or words. Ungrammatical sentences often make the reader work harder and increase the ambiguity of your statements. Furthermore, they often reflect poorly on you as a writer.

Consider the following:

---

*Memorandum*

TO: ALL UNITS
FROM: HEATING PLANT
RE: TEMPORARY ABATEMENT OF SERVICES

As you all are acutely aware, we have, for the last several weeks, been encountering less than satisfactory service with the provision of heat throughout the grounds. Repeated terminations of steam to destinations in the northwest quadrant have been particularly vexing and contributory to the production problems in that sector. We have been attempting to solve these problems relating to service through piecemeal repairs on our part. A final resolution of the situation may only be accomplished by major repairs to the central facilities and the distribution network. In an earlier memo we proposed a date for total shutdown of the system so that the requisite alterations could be done. It is manifestly clear that the suggested date is unacceptable to many of you. Therefore, no shutdown will occur on the original date in the memo.

A consensus seems to be that the dates, March 3 to 6, would present the least unacceptable timeframe for the proposed shutdown and we are proceeding with planning for those dates. While such a time may create some inconvenience at all levels of our operations, we will endeavor to complete repairs in a timely fashion to obviate further unpredicted shutdowns which are even more inconvenient and disruptive. Of course, the shutdown will create special problems for some units where provision of alternative heat sources will be necessitated. Please contact this office so that we may endeavor to mediate these difficulties through other means at our disposal. While accommodation of the consequences of this temporary disruption cannot, in all cases be achieved, we believe that most serious circumstances may be avoided with advance cooperation and planning. Please contact us for additional information.

Please, therefore, notify all concerned that there will be no heat throughout the physical facilities on March 3-6. Employees may wish to bring warm clothing and work may be rescheduled in recognition of the discomfort of no heat. We regret this inevitable situation and request your tolerance and cooperation. In the meantime, let's all hope for a warm spell during the designated timeframe.

---

We suppose that the careful reader may eventually figure out what this memo is all about. However, the reader is taken through a tortuous path of unnecessary information and even more unnecessary language of the high-falutin-high-pollutin' type, which does nothing to help us understand. We believe that the essential information of this memo could have been condensed to about one-half this length. The fancy words may impress a few readers, but most will either have to turn to their dictionaries (which is highly unlikely), will guess the meaning (more likely), or will simply ignore the memo because of the effort necessary to decipher it (most likely). Take a few minutes and rewrite this memo using clear simple language and following the suggestions we gave earlier. State the essential information at the beginning, clarify what is necessary for the readers' understanding, and conclude with a summarization of the critical points as they relate to the reader.

Type memos or print them neatly. Scrawled notes may lead to serious misunderstandings, as well as create the impression that you consider the content relatively unimportant. The form shown in earlier examples is fairly standard (if your company doesn't provide printed memo blanks). Check your spelling, and be sure that the memo is attractive in appearance.

As a current ad on T.V. says, "It's the little things that count," in creating a positive impression with others. Sloppy memos foster impressions of sloppy work all around.

Writing is an almost inevitable part of the job for engineers as they advance. Effective writing skills are dependent upon your recognition of the need to communicate effectively to accomplish your purpose with specific readers. Determine your readers and constantly review and edit your writing to address the person with whom you wish to communicate. Remember that writing reduces the possibility of feedback. Use writing where appropriate, but make every possible effort to determine whether you have been understood: seek feedback. Be concise and specific. Read your written material back to yourself to check for errors. Purchase a dictionary if you don't have one, and use it every time you are uncertain of spelling or word meaning. "Close" is not good enough when writing as a professional. Don't let the perception others have of you be the result of your laziness in writing and spelling.

## Questions and Activities

1. Gather some examples of your recent writing. Review these examples and rewrite them for greater clarity. Have you reduced the use of jargon or unfamiliar terms? Is your writing style precise and concrete? Attempt to reduce each example by 30 percent of the words used. Can you still maintain clarity with fewer words?
2. Write an abstract of an article or report you have written or have read recently. Can you summarize the material in less than 250 words, yet include the relevant material? Identify the keywords in the article that are critical for identification or retrieval of the article.
3. Look at examples of memos or letters you have written. Is it clear, right from the beginning in each, what information was contained or requested in the letter? Rewrite the examples, being sure that the first sentence or paragraph makes it clear what the reader is supposed to know or do as a result of the correspondence.
4. Rewrite the memo from the text which describes the impending heating plant shut-down. Can you rid it of redundancy, jargon, and "high-falutin" words? You should be able to reduce that memo by more than half, and still convey the same information.

## References

1. Tompkins, P. "Management qua communication in rocket research and development." *Communication Monographs*, 1977, 44, 1-26.

# CHAPTER 5

# Gaining and Sharing Information

We have already discussed several ways you can convey information to others. Oral and written reports, however, comprise only a part of our activities as communicators. Perhaps the greatest portion of our communication time is spent in direct, fact-to-face interaction with our colleagues. Our ability to maintain relationships, gather information personally, discuss information, and make decisions with others frequently determines our job effectiveness.

This chapter is designed to provide you, the reader, with some insights about developing personal relationships and sharing information. We focus here on face-to-face interaction with either one person or a small group of individuals.

Communicating with one other person—engaging in what is often called interpersonal communication—is very similar to the other forms of communication—we have discussed. The need for adaptation to the receiver is every bit as crucial in interpersonal settings as in speaking before a large group or presenting a report. Frequently (as we will discuss), there is also a great need for planning in advance of our encounters with individuals or small groups. Structuring our communication through choice of appropriate content and organizational strategies is often as beneficial here as elsewhere.

Of course, we don't mean to imply that all of your communication with others should be highly planned and structured. We suspect our lives would be mostly devoid of pleasure if we always acted from some type of plan, and always weighed every word for its impact. We need spontaneous—even uninhibited—conversation with others to sustain our friendships as well as our sanity. Few of us would be welcome as conversationalists if we *always* had some rigid structure to our interactions. Part of the benefit of face-to-face interaction is the ability to deviate from plans and explore the unanticipated.

We do, however, feel that many occasions of great importance to us professionally mandate our conscious effort to improve the communication. These purposeful interactions are the prime focus of this chapter.

While other forms of communication also have, at base, the goal of fostering understanding between two or more individuals, interpersonal communication as we use it here implies something different than those other settings. It implies a greater opportunity for mutual sharing, for equal turn-taking among communicators, and for direct, immediate feedback.

## INTERVIEWING

One situation in which interpersonal communication skills are important is the interview. Our focus in this book is on information gathering interviews, and not on employment or counseling interviews; however, much of our discussion could be equally applicable in those settings.

Information-gathering interviews comprise a large share of our communication time. All of us constantly find ourselves asking questions of our employers, managers, or fellow workers. For engineers, these interviews are often critical steps in the process of initiating, developing and implementing technological projects. Engineers often need technical advice and clarification from others. For this reason, we suggest that you should become more conscious of how you can structure your communication behavior in interviews.

Interviewing involves the establishment of a relationship, as well as the exchange of content information. A defensive climate may reduce the willingness of others to respond openly and completely. We often find the outcome of interviews in which incomplete content sharing occurred to be comments such as, "Well, I'm sorry, but since you didn't ask me that, I didn't offer the information," or, "You can find the rest of this somewhere else, I'll just point you in the right direction."

Since so much of effective communication involves an ability to put ourselves "into the shoes" of others and see their general perspectives, defensive climates often lead to an inability to correctly understand the content being transmitted. When others are defensive they may be unwilling to monitor feedback and clarify any misunderstandings, as well.

Effective interviewing for information implies the creation of supportive relationships. We create these climates through the use of supportive and confirming conversation, as well as a willingness (at times), to take a "one down" or complementary relationship with others.

Effective interviewing involves planning as well as the adoption of a supportive relationship style. The content, even in positive relational climates, must be determined, just as for an effective oral or written report. To begin, an interview requires analysis.

Just as you would not begin a technological project without a problem analysis, neither should an interview be conducted without this preparation. As you begin your problem analysis for a project, it often becomes apparent that more information is needed. This can lead to a specific identification of interview goals, or, the *specific purpose* of the interview.

In advance of interviews, in addition to determining your ultimate information needs, you must determine the likely areas of contribution by the interviewee related to the goal. In effect (as in reporting), you must conduct an audience analysis. Beyond determining what abilities your interviewee may have to offer, you must determine strategies that will be most likely to bring out the necessary information from the specific interviewee.

Plan your areas of questioning in relationship to what you need to know when you are through. If possible, inform the interviewee in advance of the goals of the interview, and as much about the specific questions as possible. This allows the interviewee to prepare as well. Your next step will be the development of an outline of the interview.

## DEVELOPING AND STRUCTURING QUESTIONS

Interviews range from highly unstructured, informal, and spontaneous to highly structured, formal and preplanned exchanges. In the latter case, every question to be asked is prepared in advance, then posed in order during the interview. This strategy is useful for mail surveys or other settings in which face-to-face interaction is not possible. It is generally too restricting for interpersonal interviewing, and denies the opportunity to follow up, clarify and make use of the available feedback in the face-to-face setting.

On the other hand, unplanned interviews may be sufficient for gathering small bits of concrete facts, but will rarely be useful when exploring broad, complex topics. Unplanned interviews seem to inevitably lead to forgotten questions, which necessitate later follow up and consequent delays.

Most interviews for technological purposes seem to call for some moderate structure. Prepare key questions to cover the specific subareas of your information goal and use these as an outline (as in extemporaneous speaking) for your subsequent interviews. Elaborate on the questions, rephrase them, add to them, or delete them from the interview as the actual process of the interview dictates.

Prepare yourself to ask a mixture of open-ended and closed-ended questions. Closed-ended questions are those which call for a "Yes" or "No" response, or for a selection from a limited number of possible responses, like a multiple choice exam. These questions are useful for gaining definite commitment and for final clarification of areas where you already have a basic knowledge or opinion.

Open-ended questions ask interviewees to respond in their own words and without forcing choices. They are useful as follow-up to closed-ended questions, e.g., "Why did you prefer design X over design Y?" They help to capture information about the subtle differences or uncertainties which closed-ended questions won't reveal.

Both types of questions are essential in most interviews for both specificity and breadth of information exchange. Some persons suggest that interviews ought to start with a series of simple closed-ended questions, which encourage the interviewee to begin responding to you. Follow these questions with open-ended probes, which expand on the early responses. It may also be wise to finish with closed-ended questions which seek definite, concrete conclusions to the earlier general open discussions.

## COMMUNICATING DURING THE INTERVIEW

Once preparation is completed you are ready to conduct the interview. Remember our analysis on an extemporaneous style. To make best use of this style we offer the following suggestions.

1. Seek clarification directly; if you aren't clear about something ask for more information. Request definitions and examples. Don't assume the meaning of things will be cleared up on their own later in this interview.
2. Summarize your understanding of the responses. Paraphrase the response to the interviewee as feedback about your understanding. Then specifically ask if your understanding is correct.
3. Make applications of the responses. For instance, generate examples or analyses through questions which begin, with, "So, it would work like . . .?"
4. Seek to create a dialogue on a spontaneous level when exploring areas of mutual uncertainty, or when jointly seeking new ways to arrive at answers. Don't back yourself into a rigid

pattern of question then answer, then question, then answer, all of the time.
5. Probe after each completed response. Ask, "Is that all?" or "Any other problems we should be aware of?"

## Maintaining Relationships in Interviewing

Effective interviewing involves not just the preparation of questions, but the fostering of an open interpersonal relationship for the free flow of information to take place.

Our relationships with other people don't just happen as if by magic. Our ability to get along well with others and to share information with them effectively develops through our interactions, and as a result of our communication with them. While it is easy to see beginnings and endings to every encounter or conversation with others, each conversation takes place within a context of our overall relationship. There is usually a chain of association in the past, and anticipation of future communication, which guides (and to some extent control) our conversation. In other words, how we will be understood is determined, in part, by the preconceptions we have about each other from past experience, and by our anticipation of future needs for interaction.

F.E.X. Dance portrays this process of communication through his helectical model.[1] See Figure 5-1.

This model shows the development of an individual over time. We constantly loop back on our previous experiences, even as we develop and change across time. Using the same concept of the helix, we can portray the interaction of two individuals.

**Figure 5-1 Dance helix**

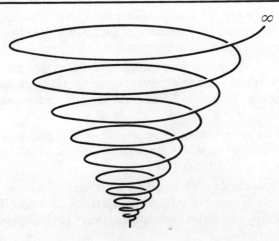

## Figure 5-2 Adapted Dance helix

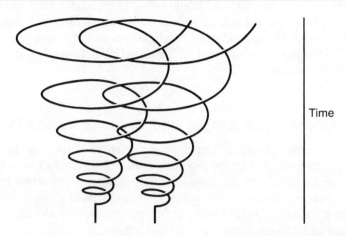

Our individual past experiences help determine our perceptions of present communication. At some point we encounter another individual and establish the foundations of our relationship. Subsequent conversations with that individual refer back to the earlier encounters to help us communicate efficiently. At any point, we may also perceive future encounters, and structure the present communication to facilitate our future encounters.

What this all indicates is that we need to be concerned with creating a good foundation in our relationships, since these early steps have such influence on the present. It also dictates that, to some degree, we need to review our relationships and previous communication experiences with others as we plan future interactions with them.

## Establishing Relationships

The research of Watzlawick, Beavin and Jackson (among others) suggests that when we communicate with others, our conversation is interpretable on two levels: content and relationship.[2] The content dimension of communication is the "what" of the conversation, the topic of interaction, and the information about that topic. The relationship dimension is the part of a message which we may perceive to carry information about the way the interactants see each other as persons, and how they feel about one another. According to many scholars, this latter dimension is critical in the creation of conditions that make communication more effective.

This approach suggests, simply, that often how you say things is as important as what you say. We have pointed toward this idea throughout the earlier parts of this book. Now we would like to share some specifics which others have suggested may significantly influence the relationship dimension of communication. We believe, particularly, that the relationship dimensions have a strong influence on our ability to communicate content, as well. If the relational implications of our communication are "wrong" to others, the content or informational impact is likely to be distorted or ignored.

## Relationship Messages and Communication

In the mid-1950s a psychologist named Timothy Leary first introduced the concept of the *interpersonal reflex*.[3] Leary's idea was that, by the way in which we communicate with others, we tend to draw forth return communication, or feedback, which reinforces our original behavior. Subsequent thinking about this reflex has led to the identification of two primary dimensions in which it works.

One dimension is that of dominance versus submission, and the other is that of love versus hate. Communicators who use a dominating style, always assuming others need to be directed and controlled, tend to receive (at least initially) submission responses from others. This leads the dominating communicator to believe that this style is accepted and effective.

In the love-hate dimension, communicators who exhibit negative relational messages implying dislike tend to draw forth similar responses from those with whom they communicate. The mechanics of this are fairly straightforward: If I indicate my dislike or distrust of you, then your responses are likely to be similar, since you perceive me as disliking you, and therefore a dislikeable person. When this occurs we start our relationship on rocky ground, distrusting one another and struggling to communicate openly and honestly.

What Leary implies is that we are responsible for the reactions we draw forth from others. We must face this responsibility, and work to create conditions which foster better communication by not placing the blame for our problems on the shoulders of those with whom we interact.

## Defensive and Supportive Communication

Another researcher, Jack Gibb, suggested that certain styles of communication lead to predictable outcomes in interpersonal and group communication.[4] Gibb labeled these two styles defensive versus supportive communication. Gibb suggested that when we communicate defensively, then others respond with their own defensive messages, communication becomes limited, and our sharing of information is reduced and/or distorted. To be effective, communication should use a supportive style.

Below are six defensive-supportive styles of communication, with a brief explanation of each.

The first pair of defensive vs. supportive communication styles is *evaluation* (defensive) vs. *description* (supportive). Gibb notes, "Speech or other behavior which appears evaluative increases defensiveness. If by expression, manner of speech, tone of voice or verbal content the sender seems to be evaluating or judging the listener, then the receiver goes on guard."[5] While we may easily sense that evaluations which are negative may make others defensive, it is perhaps less obvious that even positive evaluations may also lead to defensiveness. Even good evaluations may carry the implication that the sender is somehow given the right to pass judgment on others—an implication which may lead the receiver to resent the use of power. Any evaluation implies that we are "one-up" on others.

Gibb suggests that, where possible, we should attempt to describe our observations of behaviors without an evaluative overtone. Of course, as managers or supervisors we often must be evaluators of the performance of others. As with many things, we believe that there are appropriate times and inappropriate times for evaluation. Periodic reviews and appraisals of employee performance demand evaluation. Most of our daily communication, however, can be conducted without implications of evaluation. Employees who are constantly appraised tend, in our experience, to become dependent upon their supervisors for every move they make. They become overly cautious and non-innovative out of concern to avoid negative evaluation, and overly prone to do the tried and the true to garner positive strokes. In this situation the supervisee becomes less and less capable of independent decisions or work, and the supervisors may find that excessive demands are placed on their time.

Descriptive comments, on the other hand, can lead to supervisees and co-workers developing the ability to evaluate their own performance in reasonable ways. Description often takes the form of seeking information or perceptions from the receiver by asking non-evaluative questions. While a statement such as, "Nobody will ever accept this draft" is almost sure to create a defensive reaction, a question such as, "What do you think the response will be to this draft?" or a statement such as, "There seem to me to be two or three things we should add here," may be less negative in their outcome. Of course, even the most descriptive of words can be seen as evaluative if delivered in a sarcastic tone of voice, or with similar implications such as rolling the eyeballs up to the ceiling.

Gibb's second pairing is *control* (defensive) vs. *problem orientation* (supportive). When we constantly attempt to control the behavior and even the thinking of others through our communication (even when this control is exerted through attempts to persuade others of the correctness of our position), we can expect that receivers will respond defensively. When, however, we approach situations with a desire to *cooperatively* arrive at solutions or decisions, we can achieve better long-term relationships with our employees. Listening to others' suggestions before attempting to impose our own decisions encourages sharing of ideas that can really pay off.

*Strategy*, (defensive) as opposed to *spontaneity* (supportive), can reduce communication effectiveness as well. In essence, Gibb suggests that if we are overly calculating in planning how we will deal with others, we shut the door on idea sharing. People are adept at detecting hidden agendas in our communication, and resent our unwillingness to be forthright and responsive to their reasonable criticisms and suggestions. As an instance, this author is often accused of using "some communication technique" when conversing with his spouse. She senses (unfairly, we hope) that he is not really seeking to reach a mutual agreement, but is using a strategy to manipulate the situation. When she arrives at this perception, there is little hope of an easy resolution of the problem. An impasse is reached and the communication comes to a screeching halt. Being manipulative and using "techniques" may help us in the short run, but are rarely effective strategies for managing relationships (such as those on-the-job) which must last for longer periods.

Having stated that strategies are defensiveness producing, we must now admit to having a problem: We aren't sure how to teach someone to be spontaneous. Attempts to be spontaneous can themselves be a form of strategy. We often find ourselves very uncomfortable with colleagues who have mastered listening techniques that seem to foster spontaneity, since we suspect that they are just techniques. Our own feeling is that spontaneity is the natural result of the avoidance of the other defensiveness-creating communication factors identified by Gibb. When we are less suspicious and defensive of others, then we are better able to open ourselves to new (and potentially more risky) ways of communicating with them; ways which are dictated by the situation and the relation, rather than our desire to force our own positions on others.

In his fourth pairing, Gibb further suggests that we create defensiveness when we adopt attitudes of *neutrality* (defensive) toward others, rather than exhibiting what he terms *empathy* (supportive). Neutrality should not be confused with non-evaluativeness. It is communication which implies that we neither care about other people, nor are we willing to put ourselves in their position and try to understand them. Never revealing how we feel, and denying the importance or legitimacy of others' feelings or opinions (as though they shouldn't feel the way they do), will lead to distrust as well as reduced and distorted communication.

The fifth pair of defensive and supportive styles is *superiority* (defensive) vs. *equality* (supportive). We think that this is one of the easier ones to identify with. Even when we are subordinate to others in organizations, none of us seem to enjoy having that fact rubbed in our faces. Sometimes bosses have to act like a boss, but often they can minimize the put-downs in communication. A willingness to listen, allowing others some control of situations and conversation, sharing responsibility and credit, and allowing others to influence you helps achieve trust and honest productive communication. Many of us unconsciously exhibit superiority through our behaviors, as well as our talk. Intrusions on others' space, or denial of their privacy, touching others or patting them on the back (or other areas) when reciprocity of touch isn't allowed, and creating barriers that keep others from having access to us are examples of behaviors which are easily interpreted as reinforcement of superiority. These behaviors do nothing productive except massage your own ego, and that type of productivity is not at all useful in maintaining effective relationships which could, in turn, create better communication.

The final contrast in styles in Gibb's system is *certainty* (defensive) vs. *provisionalism* (supportive). When we approach others with a know-it-all attitude, or with the idea that we have the final answer before consultation, people rapidly become unwilling to communicate with us. We have colleagues who never seem willing to give an inch on any topic. They always have to be right in the end. These people aren't open to legitimate reasoning or problem solving approaches. Therefore, we simply avoid them where it's at all possible. When we are forced to deal with them we try to steer them toward neutral topics, like the weather. They have produced a defensive reaction in us—we perceive that it is not worth the energy to reason with them, or to engage in any conversation of substance. These people are, unfortunately, far too common. Supportive communicators, on the other hand, approach conversation and problem solving with the attitude that there is always something we can learn if we listen, that any idea can be improved if enough good minds apply themselves, that there are probably better solutions to problems than we have come up with, and that we can't force our ideas simply because they are ours. Provisionalism involves accepting our own limitations and the legitimacy of others' ideas, and being willing to change and grow. Provisionalism leads toward mutually derived solutions, which in turn may be expected to encourage everyone to carry out the decision. Decisions which we force upon others are rarely implemented with the enthusiasm and efficiency of decisions in which others have shared.

We think that you can see that many of the categories of communication and behavior offered by Gibb are highly related to each other. Working on one area alone is unlikely to achieve much in improving our ability to communicate with others effectively. To create supportive climates for good communication flow, all of these behaviors must reinforce one another.

## *Confirming and Disconfirming Communication*

In the same vein, Sieburg,[6] and Sieburg and Larson,[7] and Dance and Larson[8] suggest several types of communication behavior which are either confirming and imply acceptance of the other, or disconfirming and imply rejection of the other as a valued person.

Dance and Larson (1976) summarize these findings and suggest that, regardless of the situation in which we communicate, we are concerned with what the conversation implies about how others see us as individuals and in our relationships with others. That is, we seem to be particularly concerned with how others evaluate us, and whether the evaluations we receive match our own sense of our qualities and competencies. They note that the research indicates that whenever we receive messages which state or imply either acceptance or rejection of our self-image, we often shift our focus of attention away from the topic of conversation—the content of the message—toward the relational implications and "either or both of two objects: (1) an evaluation of [the] other, (2) evaluation of self."[9]

When we send messages which are interpreted as also carrying information about our relationships with others, or about our evaluations of others, the content—the instructions, the questions, the descriptions, the information, etc.—is more likely to be ignored or seriously distorted, since it becomes a secondary focus of the receiver. We may seriously reduce our efficiency of information sharing, particularly as it relates to the specifics of the task at hand. Our listener's task is complicated and accuracy is compromised. We believe that both acceptance and rejection messages may be a major interference in on-the-job communication, although acceptance messages may also enhance communication effectiveness.

## Rejection Messages

As we discussed earlier (in our presentation of Jack Gibb's defensive categories), negative evaluative messages may produce defensive reactions. Thus, a message which leads the receivers to believe that their self-images as worthwhile, competent individuals is rejected, may cause the receivers to respond in a number of ways that can further impede communication. A receiver may mentally note the rejection, and search for further clues that there are problems internally saying "Uhoh, am I in big trouble here or not? I'd better figure out how to defend myself." The receiver may also react by evaluating the other in a manner such as, "Who's he to say that? He's not competent to pass judgment on me! I'm tired of his attitude toward me." This not only takes attention away from the immediate conversation, but begins to erode the ability of the sender to communicate in the future. Another possible receiver response is to attack the sender by dredging up some old transgression or failure (no matter how irrelevant to the present situation) the sender has committed and say, in effect, "Oh yeah. You're not so hot, either." This type of one-upsmanship, as with the other possible responses, may lead the original sender in turn, to escalate the attacks and the defensiveness, in order to never leave the field of battle a loser. These spirals of attack and counterattack often lead to longstanding and nearly insurmountable grudges, which negate any cooperative efforts.

To summarize the outcomes of rejecting responses, it clearly appears that patterns of communication of this sort result in the deterioration of trust among communicators. Mellinger proposes that this lack of trust leads us to conceal our attitudes about issues from those who have rejected us.[10] The communication may frequently be evasive, aggressive, or overly compliant. This in turn leads to inaccurate perceptions on the part of the original mistrusted communicator. The original communicator then produces communication which reflects this inaccurate perception, all of which leads to greater loss of trust.

Several specific patterns of communication are proposed by Dance and Larson (1976) as leading to perceptions of rejection with relative frequency. Some of these are *explicit*, in that they directly reject others, while others are *implicit*, in that they reject others through more subtle cues. Below, each of these patterns is described.

**Explicit Rejection Messages.** The first explicit rejection message is, simply, a *negative evaluation of the other person*. The message states that there is something bad, or wrong, with the receiver. A statement such as, "You just aren't up to handling this project," explicitly reject another's self-image as a competent individual.

Rejection is also explicit in messages which contain a *negative evaluation of the communication content* of the other person. Here, the sender overtly denies the value or credibility of the receiver's previous messages. Statements such as, "This report isn't useful to us," or, "That kind of talk won't get us anywhere," again deny the worthiness of the receiver and destroy trust between interactants.

A third explicit message is termed *overt dismissal of the other person*, and is exemplified by a message such as, "I really don't have time right now to help you" (especially if it's clear that you do have the time).

The final explicit category is *overt dismissal of communication content of the other*. Stating that, "Your ideas aren't any use to us since you don't have the background," indicates little regard for the receiver.

While there are undoubtedly many persons who have difficulty in avoiding these overt types of rejection, we believe that they are fairly easy to reduce if you make a conscious effort to monitor your comments. The adage, "if you can't say something nice, then don't say anything," is familiar to most of us. Unfortunately, we often find ourselves in the position of violating this rule under the guise of supervising others, or guiding them. We don't expect everyone to avoid being critical of others all of the time. But, it is worthwhile to consider whether these types of *put-downs* are a regular part of your pattern of communication. You may unconsciously exhibit this pattern. As Gibb suggested, try to adopt a style which emphasizes description, rather than the overt evaluation these explicit rejection messages present.

**Implicit Rejection Messages.** While we may be able, through attention and effort, to reduce our explicit messages of rejection, implicit messages pose a greater challenge. While Dance and Larson propose several patterns which seem, with relative consistency, to lead receivers to sense rejection, there are almost infinite possibilities for receivers to somehow interpret other messages as implying rejection. What may be perceived by one person at one time as supportive and accepting, may to another person and/or at a different time be perceived as rejecting. Nonetheless, we offer the following categories as perhaps the most troublesome, and therefore most worthy of your attention.

The first category of implicit rejection messages is comprised of *interruptions*. Cutting off a speaker before he or she is done talking implies that either we are not interested in the speaker's message, that the speaker isn't worth listening to, or, most damaging, that we believe what we have to say is more important. Thus, interruptions are not merely a violation of a polite rule of conversation, but a put-down of the other person. It would appear that persons in higher positions tend to interrupt their subordinates more than the reverse. This further indicates that interruptions seem to be a subtle way of exerting and reinforcing our power over others, whether or not this power is legitimate and recognized by the other. Relating this concept to Gibb's work, (which we discussed earlier), when we interrupt we deny equality, and can expect that defensiveness and distorted communication will ensue.

A behavior called *imperviousness* is Dance and Larson's second implicit rejection category. Imperviousness describes situations where we simply fail to acknowledge or respond to the communication we receive from others. Walking past an associate without a greeting, failing to answer a direct question or return a phone call or respond to a memo are common examples of impervious, and, therefore, rejection messages. Even more common, perhaps, is the behavior of many individuals who simply make themselves inaccessible to their colleagues or subordinates. Creating barriers to communication such as a rigid appointments only structure (sometimes enforced by secretaries or receptionists who are valued for their ability to discourage anyone you don't wish to see), and never making yourself available for informal interactions with others implies that you have little or no use for them.

Yet another rejection message is often carried by an *irrelevant response*. When we acknowledge the other only by responding, but not to the topic or concern which they have raised, we deny the importance of their message and of their person. When we control the agenda, or topic of conversation, we imply that we have power and superiority. Research (Redding, 1972) suggests that we, especially when we are in the role of superior, need to allow others some control over both initiation of communication (reducing imperviousness) and the content of those conversations (reducing irrelevance).

Another rejection behavior related to irrelevance is *tangential responses*. Perhaps not as severe as irrelevant responses, tangential responses still imply that the other's topic of conversation is inferior to our own communication desires. A tangential response acknowledges the other, and briefly acknowledges the content of the other's conversation, but rapidly shifts the conversation to another topic. Statements such as, "Oh, you're still worried about that design, huh . . . well . . . uh . . . I've been thinking about our *new* project and need your inputs," suggest that the other person's concerns aren't worthwhile or even legitimate. We ask the others to drop their concerns in favor of our own.

We aren't suggesting that you, as a communicator, must always be concerned with other people's ideas at the expense of your own. We do suggest that if you exhibit a consistent pattern of denial of the others and of their concerns, then there will be a deterioration in your communication, and subsequent job effectiveness. People who experience this denial of their concerns will, we expect, become less and less likely to share even their most important worries or problems at times of crisis, since they will believe that you are not willing to really listen. Think about your communication—are you acknowledging and responding to the others and to their content of communication? Are you allowing them some control over the conversation? If not, then you are like many of us: heading for trouble in gaining and sharing information with others. Weick (1969) suggests that we are only able to understand and reduce the uncertainties of others in organizations if we don't ourselves over-structure our communication. When we assume that we know the things with which we must concern ourselves in dealing with others we shut off any chance of discovering new areas of concern or problems. This in turn leads to an inability to head off further problems before they are of such magnitude as to threaten the performance of the organization. To use a somewhat tired adage, the ability to listen to others and put aside, at least some of the time, our own concerns, is the "stitch in time, which saves nine."

## Acceptance Messages

We noted earlier that both rejection and acceptance messages can interfere with effective communication. Acceptance messages, however, offer many benefits which may outweigh the problems, provided we are aware of, and compensate for, the interference potential.

Acceptance messages, like rejection messages, cause a shift of attention away from the content or overt topic of conversation and toward the relational implications. Thus when we combine even very positive relational messages with other important content, that content may be distorted or ignored by the receiver.

On the other hand, when we are supportive of others through our acceptance of them, we develop a climate of trust, which encourages others to share information with us. When others sense that their successes are acknowledged they may be expected to increase the amount of risk they are willing to take as communicators. They become willing to share information or problems which may make them vulnerable to criticism or rejection, because they have the knowledge that the rejection is balanced by positive feedback, which reduces the importance of their shortcomings.

Moreover, because recipients of acceptance messages are more willing to reflect their honest feelings and perceptions, those who have given them acceptance derive a clearer picture, enabling more efficient communication. A supervisor who has developed a climate of trust is better able to learn. Eventually, the supervisor will be able to predict where subordinates will have difficulties, and head off the problems in advance, through planned communication of needed information and instructions. This, in turn, assists the subordinate: to do a good job, to trust the supervisor more as a person with insight and concern, and to increase risk-taking communication.

In organizations in which no one is willing to risk self-image in communication, there is likely to be greater uncertainty about how employees should perform their jobs, less sharing of creative ideas, and resultant loss of productivity coupled with increased lack of innovation. Valuable resources of the organization, the members of the company, will be lost.

Several behaviors are offered as typically increasing the feelings of acceptance by receivers. These are discussed in the following sections.

**Explicit Acceptance.** Dance and Larson suggest two types of messages which overtly state acceptance of the other person. The first of these are messages which provide *positive evaluation of the other*. The opposite of the negative evaluations discussed previously, these behaviors state that the other person is seen as exhibiting "good" characteristics. Statements such as, "You are the right person for this job," or, "We know that you are the one who has the creative strengths down here," acknowledge and reinforce the receiver.

The second explicit acceptance is *positive evaluation of the communication content* of others. Merely agreeing with others carries some acceptance message, as when we say, "That's a good report," but in true acceptance, reference is also made to the quality of the source. An example would be, "We're going to implement your plan since it's a good one. You really know how to put these things together."

**Implicit Acceptance.** There are also more subtle means of supporting or accepting others. The first communication pattern is the use of *clarifying responses.* These are statements or questions offered in response to the communication we receive. They may take the form of seeking further information about the topic ("Tell me more about this part of the design, I think you've got something!") or may be a paraphrasing of the previous statement ("So what you're saying is that we shouldn't use this material because it won't handle that stress.") These patterns indicate that you have listened to the other person and given some thought to the content, thus implying that the other person's ideas are worthwhile, and that you see him or her as a valued person.

*Expression of positive feeling* is another implicit acceptance response. When we acknowledge that the other has made us feel good, or that we have enjoyed our conversation, we reinforce a feeling of trust and let the other person know how we feel. They aren't left with the task of trying to surmise what we feel about them.

Another acceptance category is the use of *direct responses.* This one is more obvious than others, but the degree of implication it contains is often overlooked. Providing immediate and appropriate follow-ups to requests for action or questions shows others that we care about their problems and are willing to cooperate with them. Failure to respond rapidly (imperviousness) or appropriately (irrelevant or tangential responses) lead others to believe that our own concerns, or the concerns of other co-workers, come before theirs—that they are less important. To us, there is no substitute for (nor better communication strategy than) clear and rapid response to others' requests to communicate, and to the concerns raised in that communication.

Remember that *how* you say things is as critical as what you say. Any message may be interpreted as containing either positive or negative messages of evaluation, especially if they appear to be contradicted by other behaviors. A positive evaluation of another person, delivered without any eye contact, or in some other impersonal fashion, may not appear complimentary. More importantly, because relationships develop over time, and because we interpret the symbolic messages we receive within the framework of our perception of the overall relationship we have, messages which attempt to reinforce and support others may not be perceived as such when the receiver feels that our normal pattern is one of rejection. They may choose to seek the hidden rejection within or behind the acceptance. Hence, the reception of an acceptance message may start a complicated process of evaluating you, the message, and the context. This may distort both other message content and your intention to provide acceptance. The key here seems to be the consistency with which you use your acceptance (or rejection) patterns of communicating.

## Complementarity and Symmetricality in Communication

A final distinction of communication styles is that of *complemen-*

*tarity vs. symmetricality in conversation.*[11] This system proposes that we establish various types of relationships with others. In some of these we tend to try to stay *one-up* on others by taking control of the conversation, or by dominating. In these instances the others may respond with *one down*, or submission messages, in which case the relationship is a complementary one. When we refuse to respond to a one up message by submission, and instead attempt to dominate, we have a symmetrical relationship, one which implies relative equality. There are other forms of symmetricality, such as both communicators attempting to take the one down, or submission pattern. We often encounter this pattern when neither communicator wishes to make a particularly tough decision.

Neither complementarity or symmetricality is necessarily good or bad. Complementary relationships, according to Wilmot, often are valuable for the sharing of information from qualified expert to recognized subordinates (e.g. teacher/student).[12] Symmetrical relationships may be more useful in peer groups, but can degenerate into great competitiveness.

The concepts of relational messages presented in this section seem to imply that we must be concerned with *how* we communicate, as well as with *what* we communicate. If you create defensive or disconfirming environments through your communication, then the likely outcome will be a reduction in the quality of your relationship with others. Since, as the Dance Helix implies, we constantly reflect back upon earlier experiences as we continue through relationships, we believe that the creation of these negative atmospheres will seriously jeopardize our ability to survive and thrive in the larger organizational environment. At worst, we can get caught in continuous spirals of negative self-fulfilling prophecies. Our negative expectations lead us to act defensively or in a disconfirming pattern—which leads others to react defensively in a disconfirming fashion—which leads to reinforcement of our initial negative assumptions about the relationship. Few of us can afford many such spirals in our professional relationships.

Poor communication techniques lead to less information sharing, and less accurate understanding of information that is received. The communication styles and techniques we have described determine, in part, the *climate* within which you communicate now and in the future. Even when you are informally seeking small amounts of information—an explanation of new software availability, a deadline for a feasibility report, a stress factor for equipment, or whatever—your communication style will influence the person from whom you seek the information. A poor climate is one with distrust and lack of openness. It is brought on by defensive and disconfirming communication on your part, and cannot fail to make your task as an interviewer more difficult.

So far, we have discussed the importance of developing plans for interviews: determining the information sought, the questions which will focus on that information, and the ordering of the questions. We have also described several key characteristics about *how* you can communicate better to foster good relationships. Finally, it is important to recognize that effective interviewing requires an appropriate conclusion to the process.

## CONCLUDING AND FOLLOWING UP ON INTERVIEWS

Remember that conclusions are vital for effective interviews. The whole structure of an interview from start to finish can be a modification of the old public speaking organizational strategy of, "Tell them what you're going to tell them, then tell them, then tell them what you told them." For interviews, "Tell them in advance what you will ask them, ask them, then tell them what they told you."

Provide a general summarization of the interview. Indicate, where possible, how the information will now be applied and what, if any, follow-up the interviewee may expect about the results of the interview.

Acknowledge the contribution both immediately and with a subsequent memo or letter. Keep the channel open for further communication by offering a positive relationship. All too often, interviewees think of something relevant a few minutes after you walk out the door—once you discuss an area the interviewee tends to be sensitized to that area. A follow up letter may be just enough to encourage transmission of this new information, rather than encouraging the attitude, "Oh well, she's gone now. She'll figure it out sooner or later on her own."

The follow-up letter contains information about your application of the information that allows for correction of misinterpretations. Acknowledgement confirms the interviewee's relationship to you, and creates a positive spiral for future interactions.

## LISTENING

For all interviewing (as well as any other interpersonal contacts), an essential skill is listening. Recently, we've become more and more aware of how we often don't listen effectively. In Chapter 1 we discussed how communication involves sharing meaning. The listener is as active in communication as the speaker. Without the listener acting on the information transmitted, no communication has taken place. Poor listening inevitably leads to poor communication, and costs us dearly in our professional performance. Too often, we hear engineers complain that others don't listen to them well. Almost as often, engineers tell us they have trouble remembering all the important information *they* hear.

Below are some suggestions for listening effectively. Many of these are discussed in greater detail by Carl Weaver, as well as other scholars studying the listening process.[13]

I. **Improving your capacity to listen**

   A. As you listen, focus on listening, not on your next statement.
   B. Try to determine the general *frame* of the speaker from vocal and nonvocal cues, as well as your knowledge of the person's background and experience.
   C. When words or terms seem ambiguous to you seek to establish their definition.
   D. Consciously summarize the statements as you go.
   E. Reflect the statements back to the speaker through paraphrasing, etc.
   F. Remember that confirming responses, either vocal or nonvocal, encourage the speaker to offer more.
   G. Listen for ideas, not for style.
   H. Try to connect the unfamiliar with the familiar; create your own analogies, then check them out.
   I. Take notes while listening if possible or necessary, but not at the expense of paying attention.
   J. Seek the cause of inconsistencies: is it you, or the sender?
   K. Write yourself a note or a memo following the conversation, speech, interview, etc; summarizing to provide your own reinforcement—send the memo to the source for verification and correction if necessary.
   L. Follow up promptly.

II. **The most important rule may be to remember to monitor both your own communication and that of others constantly on all levels.**

   A. People value others who have "something to say" when they speak.
   B. People value others who listen carefully and respond to the communication content and the immediate relationship, rather than to their own hidden agendas.

## SHARING INFORMATION IN GROUPS

The use of small groups is widespread among engineers. Often engineers are part of coordinated teams of specialists who interact constantly to solve technological problems. Many engineers complain to us about inefficient information sharing and decision making of the groups they are a part of.

They are not alone in this complaint. Fortunately, there are some ways in which group communication may be improved.

To begin this section we feel that we should review some of the reasons for using groups, since these reasons often influence how we should communicate. Beebe and Masterson (1981) suggest that there are several advantages and disadvantages to using groups.[14] These are summarized in the following box:

---

**Advantages**
1. Groups have greater resources of knowledge and information.
2. Groups can employ greater numbers of problem solving techniques.
3. Working in groups fosters improved learning and comprehension of ideas and suggestions discussed.
4. Members' satisfaction with the group decision increases because they participate in the problem solving process.
5. Group members gain a better understanding of themselves as they interact with others.

**Disadvantages**
1. Group members may pressure others to conform to the majority opinion.
2. An individual group member may dominate the discussion.
3. Some group members may rely too much upon others to get the job done.
4. Solving a problem takes longer as a group than as an individual.

---

To us, two of the most important of these advantages are the increase in understanding and commitment to decisions groups foster. Advantages are diminished, however, if we don't structure our communication properly. Planning the communication helps guarantee all of the advantages.

So, in either large or small groups, the maintenance of relationships—attending to relational implications of our communicative behavior—is as important as the content of the communication. To be effective in groups we generally need to maintain a supportive, confirming climate. This does not mean that groups should avoid all conflict, a topic we discuss later in this section. In this section we are primarily concerned with somewhat structured group interaction, rather than the casual conversation carried on continually. Be aware, however, that this continual interaction nonetheless forms the foundation of the formal group interaction. Sound maintenance of interpersonal relationships may well determine group effectiveness.

## BEGINNING GROUP INTERACTION AND PROBLEM SOLVING

Group interaction requires problem analysis, just as reporting information or structuring interviews. Successful groups usually address this planning on two levels.

To solve a technological problem the group needs to analyze the problem area. To do so effectively, the problem analysis must involve the entire group through communication with one another. In a sense, the group needs to conduct a second analysis of the communication related to the problem. The group needs to determine what communication strategies will best enable it to arrive at a satisfactory decision.

In our limited space, we cannot identify all of the possible problem solving strategies a group may adopt. We recommend *Between You and Me: The Professional's Guide to Interpersonal Communication*, by Robert Hopper, for further suggestions. For our part, we offer some specific strategies we feel fit the needs of engineers.

We have already stated that all members of the group should be involved in problem analysis. Groups that don't do well in problem solving often perform poorly because they immediately offer solutions before everyone agrees on what the problem is. Members of the group may be proposing solutions to their individual interpretations of the problem area. The group lacks focus, and may tend toward selection of those solutions which are most forcefully presented by powerful members of the group. Such groups are solving problems on the basis of personality, rather than on the basis of logical technological criteria. Without agreement on a direction in which the group may work together, the group process seems unlikely to provide many advantages.

Problem analysis cannot be delegated to one individual in the group on the assumption that a presentation of one person's findings represents the final word on the problem. We suggest that the subareas of concern be divided among the "team" members, and that they subsequently interact directly with one another to consolidate their analysis and determine an acceptable problem statement. Interpretation of the problem area by each individual should be explicity checked out: ask one another to summarize understanding. Clarification of misunderstanding at the outset of problem solving is critical.

In one instance we have encountered, a group of engineers attempted to solve the problem of lagging productivity in an assembly process. Two of the engineers each offered a solution to the problem immediately after the meeting convened. One suggested some new equipment, while the other suggested new training techniques and incentives for the workers on the line. Each of the engineers couldn't understand why the other so adamantly refused to accept the "logical" solution, and kept advocating a "bonehead" idea. No one, at first, could understand how so much hostility and conflict could result from two straightforward solutions presented by two such straightforward and nice guys.

They didn't realize that each was addressing a fundamentally different conceptualization of the problem. One engineer saw a mechanical and technical problem, while the other saw a human resource problem. Each of their solutions was appropriate for the problem as they saw it, but of course, didn't fit the other problem.

Fortunately, after considerable wrangling, another member of the group suggested that they all take a look at the problem in a systematic fashion. Careful analysis of the problem revealed that a technical problem was indeed the critical cause of the current problem.

We would bet that if they had continued on their initial path of arguing over the two solutions, the group would not have arrived at a relatively simple, low-cost and effective solution to the current problem. Several alternatives could have emerged. The group could have split over which solution was best and then decided by majority rule which to accept. This could have left the losers disgruntled and uncommitted to further support of the solution.

A second possibility is that the group could have tired of the dispute and sought a compromise. Such a compromise could have proven costly and wasteful—it would have solved the problem, but it would have done much more than was needed. Applying a problem focus to the interaction helped this group solve their problem in a way which they all finally agreed was the proper fashion. In short, it led to an effective consensus solution, which all involved supported enthusiastically.

As any group progresses through stages of its attack on a problem, communication must be maintained. The separate stages of a project usually involve several persons working independently, then coming together at the end to share results or conclusions and to plan subsequent steps. This form of sharing may not be frequent enough. To solve this, build in more frequent communication about your activities.

It may not be necessary to gather the group together face-to-face constantly, but you can use memos or progress reports on a regular basis to inform others of how your separate subprojects are moving along. This may help avoid unnecessary duplication, and enable you to discover misinterpretation of instructions or the problem analysis before large amounts of time are wasted. In the final chapter of this book we provide a brief description of one communication system for constantly coordinating actions of large numbers of groups. Such systems can be equally effective when applied to the individual members of small groups. Of course, the small group can use face-to-face interaction more frequently and with greater success than large groups, but only if the interaction follows some structure and planning.

There are many means by which a final decision may be reached through group interaction. Perhaps the simplest of these is to vote on proposed solutions. A majority vote determines the adopted solution. Unfortunately, majority rule creates some problems. First, it often leads to choosing only from available solutions rather than finding new or combining old solutions to best advantage. It tends to lead to either/or selections: choosing either Plan A or Plan B, rather than creating a better Plan C.

Second, majority rule creates losers. Unless the vote for a particular solution is unanimous, some members of the group are forced to act on a solution they didn't support. You may recall that one of the advantages of group decisions is that members are more committed to the decisions. When there are losers this advantage is lost, at least for the outvoted members.

Good group problem solving involves efforts to arrive at consensus decisions; ones which the *whole* group believes to be acceptable. While majority rule may be necessary in some cases (such as when time is short), groups are generally better off if they strive to develop solutions to problems which all the members agree upon. Consensus decisions come from developing solutions which satisfy everyone, not ones which represent compromises that leave everyone dissatisfied with the outcome.

To structure communication to help assure consensus, it is often useful to utilize specific problem solving techniques which lead the group toward unanimity. One such technique, the *Reflective Thinking Format*, is included later in this chapter. (There are many others available in larger texts on group processes and communication.)

These techniques, in some way, force problem analysis early in the group process. They also create a structure for managing conflict. Note that we use the term *management—not resolution—*of conflict. Not all conflicts can be resolved, but most are manageable if communication is structured effectively.

# APPLICATION: THE REFLECTIVE THINKING FORMAT

One system for structuring group interaction, which we believe coincides with the concepts we have introduced, is called the Reflective Thinking Format (Beebe & Masterson, 1982[15]). Based on John Dewey's conceptualization of how we systematically solve problems individually, it can help a group arrive at a sound consensus decision. We also like it because it is basically the same as the problem-solution report organizational strategy we covered in Chapter 3. Thus, when this group strategy is used, each step also represents a section of a technical report. This simplifies the report process, and helps assure organization of both the group process and the report.

### Step One

**Limit and Define the Problem.** As mentioned, a critical first step assures that you have reached agreement and understanding, which avoids tangential behaviors by group members. Locate the problem area and the specific concern of the group. Try to word your problem as a question; e.g., "What can be done to reduce drag?"

### Step Two

**Analyze the Problem.** Determine the fundamental and immediate causes of the problem. Decide whether fundamental or immediate causes are most relevant for a solution. Establish the specific criteria by which a solution should be evaluated. What should a solution do? What limitations must be imposed on the solution? Which criteria are most important, and how much more important are they?

### Step Three

**List Possible Solutions.** Conduct research or brainstorm to determine what solutions are currently available or may be developed to address the problem. Testing of the solutions on a trial basis may be appropriate here. Encourage creative and freewheeling generation of ideas and evaluate *later*, not during the process of generation.

### Step Four

**Determine the Best Solution.** Apply the criteria you have developed earlier to evaluate each of the possible solutions. Select the solution which best meets the criteria. If possible, combine solutions to meet most, if not all, of the criteria.

## Step Five

**Plan Implementation of the Solution.** Determine the procedures needed to implement the solution you have chosen. Delegate responsibilities for parts of the implementation. Decide on how and when to review the effectiveness of the solution.

Let's follow the management of Midland Foundry through the process of group problem solving and decision making. Midland, a relatively large foundry producing precision castings, is constantly concerned with reducing *scrap*, or defective castings. Problems in production lead to higher overhead and delays in servicing customer accounts; both serious problems in an industry facing more and more foreign competition. Lately, the percentage of scrap had increased slightly—enough that management was vitally concerned. Managers and production supervisors gathered together to determine what could and should be done.

Immediately after the meeting convened, one of the engineers present suggested that the outmoded equipment needed replacement and that was all there was to it! The financial manager immediately offered a detailed description of the company's financial condition, current interest rates, and the effects of capitalizing the new equipment that was needed.

At this point, Tom Carroll, the plant manager and an engineer, interrupted and suggested that they were jumping to conclusions. He noted they were putting forth a solution without looking at the problem. Tom encouraged the group to systematically examine the situation. He commented that, while there was no doubt that the equipment was old, maybe their current problem had nothing to do with that situation. After all, the equipment was "old before, but scrap wasn't this high. What's happening now seems like the place to look."

Tom lead a discussion of the specific nature of the problem. After talking with each of the concerned members of the group, the specific location of the problem was determined—a stage in the process was pinpointed. Having located where the problem resided, the group tried to find the causes. Gradually, through the discussion, a picture emerged. Jim Thompson, formerly a supervisor on that part of the process, said that he and his former workers had always had some difficulty with the equipment, but adjusted to the problems rapidly as they occurred. Tom asked Carl Biggers, the current supervisor, whether they had the same problems and how they adjusted to the difficulty. Carl said he thought they did a "good job down there," but, of course, two of the "old men" had taken early retirement right after Jim left. Their replacements were good workers, but maybe they didn't sense the variations in the equipment. Carl admitted, reluctantly, that he had trouble explaining it to them.

After more discussion, they were able to agree that, while the machinery was at fault, inexperience and perhaps insufficient training were the causes of the recent change in quality of production.

One of the engineers began to offer a solution, but Tom stopped him in midsentence saying, "What does the solution have to do?" One member suggested that the solution had to reduce scrap by at least 50 percent. The accountant offered that it had to fit within specific economic guidelines. Others offered criteria of time constraints and disruption of other activities. The full set of criteria included all these factors, with the understanding that preference should go to solutions which postponed extensive capital investment until the overall modernization of the plant could be accomplished. Thus, the various criteria were given a priority, or weighting.

Working together the group proposed several alternative solutions. One was to purchase some new equipment. Another was to simply wait and do nothing—with experience the workers would learn to reduce the problem. A third suggestion was the retraining of the workers. A fourth was the modification/repair of the unreliable equipment.

Applying the criteria to the solutions, no one emerged as a clear-cut winner, but all could agree that retraining of the workers best fit the criteria, for now.

Mechanics of the implementation were worked on next. Someone suggested that perhaps they could either involve Jim, the former supervisor, in the training, or, if that was impossible, lure one of the retired experts back as a temporary consultant to work with the men for a while. Dates were established to review the effectiveness of the solution and to determine if other action would be required. In the meantime, the group continued its planning for the replacement of the troublesome equipment, as well as extensive plant modernization.

Analyzing this case, several critical steps deserve recognition. The group, after a false start, shifted to a problem orientation. Causes of that problem were discussed and determined in a manner which enabled greater possibilities for solutions. While one fundamental cause remained, other immediate causes emerged from the systematic investigation. Criteria were established which everyone understood and agreed on. Given these criteria and the solutions which were offered, the emergence of one solution was nearly automatic and free of discord. From this process a spirit of cooperation emerged, with better understanding on the part of the group of the difficulty. They were willing to work together on subsequent projects.

The solution they proposed proved fairly effective, but never reduced scrap by the desired 50 percent. Nonetheless, when the review of the solution took place, the group was satisfied that they could do no more until the plant was modernized.

Structuring of group decision making can't always guarantee highly successful solutions. It does appear to increase the probability of this success. Moreover, the benefits of reduced dysfunctional conflict, increased understanding, and increased commitment make such structuring highly desirable. Unstructured communication in groups can lead to a sense that too much time is spent in meetings where everybody talks, nobody listens, and nothing ever gets done.

## CONFLICT

Conflict has earned itself a bad name in organizations. We tend to feel that conflict leads to disruption and poor problem solving. In fact, conflict serves vital functions in groups.

Conflict helps us to gain an understanding of the attitudes others hold, as well as the strength of these attitudes. If conflict is always suppressed, then we tend to be cautious in our interactions and in the ideas we are willing to present. Research by Johnson indicates that when organizations allow little conflict to occur, there is a tendency to develop only tried and true—rather than innovative—solutions to problems.[16]

We aren't advocating that you, as group members, ought to intentionally scream, kick, bite, or strangle other members of the group as a means to improve the quality of solutions. We do believe that some disagreement is legitimate, and that when leaders or other members of a group deny these disagreements, the benefits of groups are diminished.

Conflict is almost inevitable. However, it best serves the group if it occurs in early phases of problem solving, rather than when a solution is finally selected. If conflict is not present early in interaction, the pressure of group members' disagreements may build and erupt at the end, destroying group commitment to the decisions. When conflict occurs over the definition of the problem, or over criteria by which solutions will be judged, it often brings forth useful input, which strengthens the group's decisions and helps assure consensus without conflict later.

Some researchers differentiate two types of conflict: substantive and affective[17]. Substantive conflict is characterized by disagreements over the issues the group is attempting to solve. This disagreement may focus on interpretations of facts, policy to follow, acceptable criteria, or other factors of the problem. Affective conflicts concern the interpersonal relationships and feelings group members have toward other members.

Conflict over the issues may be particularly helpful in bringing out needed information about group members' ideas and understanding of the problems. Guetzkow and Gyr suggested that groups experiencing substantive conflict arrive at consensus through continual discussion of the issue, and the incorporation of all available information to manage the conflict.

When affective conflict occurs, consensus can be reached only after shifting the topic of discussion from the issue generating the affective responses to simpler issues, upon which the group can agree. Then go back to the original problem area.

Earlier in this chapter we discussed confirming vs. disconfirming responses, as well as complementary and symmetrical relationships. We suspect a relationship exists between disconfirming or one-up behaviors and our perceptions of affective conflict. When the group interaction produces disagreement over content issues, then conflict may not be a problem. However, when relational messages of evaluation of others are present, then conflict may indeed be present and dysfunctional. When disconfirming communication becomes prevalent, the best strategy is to attempt to shift the group focus toward other issues, since the disconfirming, dominant, defensive communication tends to create similar patterns in a spiraling process.

In a nutshell, we suggest that even in heated arguments about ideas and policies, keep attempting to bring in more information from all members. Air the ideas and attempt to find solutions which allow all sides of the argument to win. This doesn't necessarily mean adopting a compromising "I'll give up this if you give up that" approach. Rather, attempt to create new approaches that give all sides what they want. When disagreement focuses on personalities rather than issues, participants' focus shifts away from content. Effective problem solving in these situations is unlikely. You may create a more positive environment through your own confirming behavior or by emphasizing areas of agreement.

## Questions and Activities

1. Develop a plan for an interview to gather information related to a current project. Write a purpose, then develop an outline of key questions.
2. If possible, and with the permission of those with whom you communicate, record your conversation during an interview or group meeting. Can you detect examples of defensive or disconfirming responses which you or others made? Can you detect any pattern to these responses? Do they seem to follow and (in turn) generate similar categories of communication?
3. Think of a time recently when you were unhappy with the quality of performance of a subordinate or co-worker. Were you descriptive or evaluative in your response (or would you have been either had you chosen to say anything)? Try to develop a set of descriptive comments you *could* have made.
4. Think of the last time you sought information from a colleague. Did you attempt to structure the interview? Did you follow up the interview as suggested in the book? How would you go about assuring that the information you received was complete and accurate? How would you

structure the interaction to assure that the channels of communication remain *open*?
5. Interview a colleague or friend following the suggestions for effective listening. Consciously try to provide confirming responses, including clarification responses.
6. Analyze the last meeting in which you took part. Can you detect any specific structuring to the meeting? How was communication limited and directed? Were solutions offered before problem analysis? Were criteria agreed upon before evaluation of solutions? Was consensus achieved? Did members leave the meeting motivated to act on the solution? Why or why not?

## References

1. Dance, F. E. X. *Human communication theory: Original essays.* New York: Holt, Rinehart and Winston, 1967.
2. Watzlawick, P., Beavin, J. & Jackson, D. *Pragmatics of human communication.* New York: W. W. Norton & Co., 1967.
3. Leary, T. *Interpersonal diagnosis of personality.* New York: Ronald, 1957.
4. Gibb, J. R. "Defensive communication." *Journal of Communication,* 1961, 11, 141-148.
5. Gibb, J. 1974, p. 328
6. Sieburg, E. *Interpersonal communication: A paradigm for conceptualization and measurement.* San Diego: United States International University, 1975.
7. Sieburg, E. and Larson, C. E. "Dimensions of interpersonal response." Paper presented at the International Communication Association Convention, Phoenix, 1971.
8. Dance, F. E. X. & Larson, C. E. *The functions of human communication: A theoretical approach.* New York: Holt, Rinehart and Winston, 1976.
9. Ibid. p. 78.
10. Mellinger, G. D. "Interpersonal truth as a factor in communication." *Journal of Abnormal and Social Psychology,* 1956, *52,* 304-309.
11. Watzlawick, P., Beavin, J, and Jackson, D. *The pragmatics of human communication.*
12. Wilmot, W. W. *Dyadic communication,* 2nd ed. Reading, Mass.: Addison-Wesley, 1979.
13. Weaver, C. *Human listening, processes and behavior.* Indianapolis: The Bobbs-Merrill Co., 1972.
14. Beebe, S. A. & Masterson, J. T. *Communicating in small groups: Principles and practices,* Glenview, Illinois: Scott, Foresman, 1982.
15. Ibid.
16. Johnson, R. J. "Conflict avoidance through acceptable decisions." *Human Relations,* 1974, *27,* 71-82.
17. Guetzkow, H. and Gyr, J. "An analysis of conflict in decision making groups." *Human Relations,* 1954, *7,* 367-381.

CHAPTER 6

# Communication in the Technological Organization

Thus far we have discussed communication as it occurs in a number of specific settings. This chapter is about communication in organizations, where all of the previously described settings and forms of communication come together. Technological organizations are often complex and demand not only the previously described forms of communication, but others as well.

## IMPORTANCE OF COMMUNICATION IN THE TECHNOLOGICAL ORGANIZATION

In Chapter 1 we proposed that the ability to communicate effectively was essential for survival and advancement in organizations. As an engineer advances (on the basis of technical skills) there is an increased demand for communication skills. Engineers, as they move up, generally find themselves managing more and more. This requires greater communication, both in quantity and quality. Additionally, upward moving engineers frequently find themselves representing their organizations before the public, as well as to client organizations.

To be effective, technological communicators must ultimately coordinate all of the skills of communication, not just one or two. While early in a career the ability to maintain interpersonal relationships and to write an occasional report may suffice, at advanced levels there is often increased pressure for presentational speaking, group work, interviewing, and other forms of communication.

To be effective, you must also learn not only *how to communicate*, but also *how to manage communication* of your organization (or the parts which you manage). More and more, managers in all fields are becoming aware that, like all other aspects of organizations, communication in the organization cannot be left much to chance. To some extent, communication can and must be controlled, as are the allocation of personnel, the acquisition of raw materials, the assurance of quality control, the marketing of products, the scheduling of production, and myriad other tasks. The reason is quite simple: None of these tasks can be accomplished without communication, either in the organization or to the outside environment. If communication is inefficient, the organization will suffer as a result.

Engineers are familiar with the concept of *systems* as a result of their training and background. Therefore, we needn't dwell on the nature of systems in general, except to emphasize the interrelatedness of all the components. Communication is itself a system which links the components of the organization system. Disruptions of one sector of the organization inevitably affect other sectors. Communication can reduce the negative effects of disruption or change when messages flow efficiently between related components. Similarly, if one part of the communication system is disrupted or inefficient, there is an inevitable effect on other components of the communication structure. A broken telephone means added burdens of communication in other channels. A person who writes illegible or incomprehensible memos places an increased communication burden on others who must solicit clarification. Effective managers monitor and control the elements of the communication system to reduce system strain and inefficiency.

To be effective in the technological organization engineers must learn to reorganize and utilize the existing communication systems, even if these systems are seriously deficient, or in need of management. It does little good to sit and grumble about the inability of the system to function properly when changes may take a long time to implement, or where solutions to the problems are not easily identified and implemented. You must continue to function within the existing environment, or else risk isolation and poor personal performance.

Communication systems don't always function as planned. Informal patterns and short cuts develop, which may facilitate or impede the flow of information within the technological organization. The wise engineer studies the informal patterns in order to determine the most efficient and the most personally and professionally advantageous channels through which to direct communication.

In keeping with a systems perspective on communication in the organization, we will first discuss how the organization may interact across its external boundaries. The individual engineer may frequently be called upon to provide the bridge to the outside.

## INTERFACING WITH THE PUBLIC

More and more businesses (especially technological ones) have grown to realize the importance of presenting a controlled image to the public. Most large corporations employ both advertising and public relations firms to help enhance their images. These agencies place advertisements, generate coverage of significant events by news agencies, and provide printed material explaining the organization to outsiders. This type of external contact has been carried on for years. Now, however, many organizations have shifted their emphasis to more personal forms of representation to the public.

These organizations recognize that advertising and press coverage don't always create the positive image they desire. For one thing, a sense of the *people* who *are* the organization is often absent. Technological organizations, especially, often seemed detached and impersonal to the public, perhaps because their work is sometimes shrouded in secrecy, or is incomprehensible to the average citizen. Yet, the members of the organizations—the engineers and technologists—are very human; they are the neighbors, and the members of the congregation and the P.T.A., just like non-technologists.

Therefore, engineers may be much better "goodwill ambassadors" for the organization than impersonal advertisements. For this reason, many organizations now encourage their employees to *actively* represent the company to the public. These organizations establish speakers bureaus, or otherwise provide for employers to appear before the public to explain and represent the company.[1,2] These employees appear at meetings of citizens groups, fraternal organizations, schools, or anywhere there is a request.

Recently, the Milwaukee Metropolitan Sewerage District purchased a substantial block of time simultaneously on the major television stations in the city. Their program was an attempt to explain to the taxpayers how their money would be spent for the massive modernization of the sewage treatment system. Various components of the project were described in relatively non-technical terms for the general audience—or efforts were made at this. The entire production, which could have done much to dispel fears and misunderstanding, wasn't as effective as it could have been. In fact, we would bet that people throughout the area dusted off cribbage boards or got to know their kids again after the first few minutes. The program was disjointed, repetitive, and often unclear. A solid attempt, aimed in the right direction, failed because the individual skills of the communicators, as well as the overall management of the TV production, were inadequate. Nonetheless, management of the district should be applauded for its recognition of the need for communication to the public about projects which affect the public. We believe that the recognition of this necessity will result in a growing need for engineers to improve their public communication, and to manage the flow of information to an involved population.

These appearances can help narrow the gap of misunderstanding and distrust between the company and the public. The *personal* communication is critical. As an engineer you may have the opportunity to engage in these appearances. The organization may go so far as to provide you with a professionally scripted speech. (Speech writers may make considerable money preparing these.) This scripting may be necessary to help control the company image, and assure consistency in presenting company policy. In this manner these presentations represent an appropriate use of manuscript speaking. However, if you use such a script, remember that the personal contact is the essential part of your public relations mission. Prepare yourself adequately to present the manuscript in a conversational *personal* fashion. Don't read the script in such a way that the audience is left with the impression that the company might as well have sent a cassette recording to be played on the public address system. Respond and adapt to your audience to the extent that the requirements for uniformity of presentation allow.

Many organizations now train a wide variety of employees, and not just sales and marketing personnel, to provide representation. (You may even be reading this book as the result of your company's desire to develop your skills for such purposes.) Many such organizations seriously consider their employees' contributions as representatives when determining promotions, raises and other rewards. Therefore, participation not only enhances the company goals, but your personal goals as well. It allows you an opportunity to stand out among your peers in a positive fashion. Set your anxieties aside and take the opportunity if available. You may find you enjoy this new challenge in your professional career.

The organization may use this presentation for other purposes, as well. As a personal envoy to the public, you may also serve as the company's "ears." Your ability to perceive the feedback from audiences can enable you to help the organization refine its presentations and other public relations efforts. You may help your organization considerably if you prepare a report about your appearances summarizing audience response, the types of questions asked, and the information which appeared to have the greatest impact on your audiences.

Spanning the boundary to the public personally creates the type of true *communication* vital to an organization's successful solicitation of public support for its projects. Growing accountability for technological progress mandates this involvement with the public. Citizen hostility toward technological progress can often be reduced through the development of a sense of participation in that change, just as participation in small group decision making increases commitment. To gain this advantage, though, you and your organization must be truly committed to communicating and responding to the public. Inputs to the communication system of the organization, as well as outputs, are vital.

## INTERFACING WITH THE TECHNOLOGICAL ENVIRONMENT

Communicating with the general public isn't the only form of "boundary spanning" in which technological organization employees engage. Such companies are dependent upon the flow of new inputs from scientific research, as well as from other technological organizations. Few companies can survive in fast-changing technical environments without importing the raw material of ideas. The success of many new high technology firms appears to be closely related to their ability to create strong ties to research institutions (such as major universities).

Even as we write this, the University of Wisconsin system is reviewing policies to facilitate greater interaction by faculty with private industry. Industry, in time, needs to create means of maximizing this interaction through personal contacts.

Rosenbloom and Wolek estimated that in upwards of 70 percent of the industrial applications of scientific innovations, the source of information about the innovation was personal contacts.[3] Allen notes that even the personal contact is often ineffective because, "the average technologist cannot communicate effectively with outsiders."[4] Technologists, typically, do not read large amounts of research reports and do not effectively incorporate ideas from outside the organization.[5] This may lead to severe lags in implementation of available—but undiscovered—technology. We hope that the suggestions we have offered and the insights you have gained about interviewing may help you to bridge the gap.

Allen[6,7] has shown that, in the absence of effective communication skills on the part of most technologists, a key role has developed in technological organizations: the *technological gatekeeper*. It is through this person that new information enters the organization. This person is generally better integrated into the formal and informal networks of the field, can define these inputs for colleagues, and is generally better on a number of job performance criteria. When the gatekeeper is used as the source of input to the organization, there is a marked improvement in the quality of implementation. Clearly, from their research, a person with strong oral skills is critical to the functioning of the organization. Additionally, these skills may contribute to individual success. Technologists who exhibit these skills are often recognized for their contributions to the organization and rewarded. They serve as technological gatekeepers bringing needed information into the system.

## INTERFACING WITHIN THE TECHNOLOGICAL ORGANIZATION

Many, if not most, technological organizations are complex multi-divisional systems. The success of these organizations depends, in large part, on the ability to communicate effectively between these separate divisions. Clearly, in an interdependent system, the right hand must know what the left hand is doing.

Effective communication in such companies requires the creation and utilization of patterns and forms of communication which facilitate the timely flow of messages. Brillouin differentiates two types of information in the organizational environment: absolute and distributed.[8]

Absolute information is all the knowledge which the organization has available within its boundaries. It is the information which individual members have, or which is stored in files or elsewhere.

Distributed information is the knowledge which is disbursed among many persons in the organizational environment.

Sometimes information necessary for the functioning of an organization is known by only one person, or is filed away where no one remembers it exists. One of the key problems engineers face is identifying the absolute information and distributing this information throughout the company as needed.

Many organizations have developed sophisticated computer-assisted data management and retrieval systems to help distribute available or absolute information. Many of these systems also provide a bridge to outside information sources as well. An engineer seeking information in a particular technological or managerial area seeks this stored material, retrieves it, and then uses the information. Not all absolute information is available in data storage systems, however. Much essential information may only be acquired through personal contact and communication with individuals. This information is distributed through the organization through communication "networks."

Networks are the patterns through which information typically flows. For the organization to survive, the networks must distribute communication efficiently among subsystems and individuals throughout the system. Organizations usually develop both formal and informal networks of communication.[9]

Formal networks are those designed explicitly by management. These networks are established through required communication, such as daily reports, requisitions, planning documents, etc., which must be sent to specific units or individuals either on a regular basis or when activities are performed. In effect, management dictates who must communicate with one another and when. Matrix organizations particularly are structured in this way. The organization may also restrict communication by requiring communication among units only through the chain of command. These requirements often help upper management by keeping them informed of the activities of subordinates.

Quite frequently, these designed, formal networks prove inadequate to assure distribution of necessary information. The design may be outdated as a result of shifts in company goals or structure. Often, these formal networks are inadequate because some individual members are unable or unwilling to communicate information to other members or units. They may block communication because they wish to preserve control of their divisions, and believe that as long as other persons are uninformed about their activities then little criticism will result. They choose to keep a low profile on the assumption that "no news is good news" to upper management. We believe that such an attitude is short-sighted and destructive, both organizationally and personally.

Individuals may also block distribution of information because they have not developed relationships with other individuals which allow open sharing of information. In Chapter 5 we indicated the importance of developing good relationships through communication, so we hope you may avoid this problem personally. The fact remains that some individuals, for whatever reasons, become "isolates" in the communication network. They don't receive information, nor do they pass on information. While such an isolate may pose fewer problems if situated at the bottom of the organization totem pole, when these individuals are in higher positions isolation affects all who work under them. These subordinates must seek alternative informal networks of communication to gather input.

Many organizations now attempt to discover these informal networks of communication (or the *grapevines*, as they sometimes are called)—not to destroy them, but to discover why and where there are blockages in the formal networks. Having identified these bottlenecks in the information flow, they can attempt to correct the problem. They may do so by establishing new specific reporting requirements, or, in extreme cases, by removing the persons who are unable to maintain the communication relationships, and replacing these persons with more skilled communicators.

Management may also be able to discover those individuals who are effective in linking units of the organization together. These are the persons who interact well with members of other units. Often their interaction is informal, but provides for the essential distribution of information. These individuals may eventually be rewarded through promotion to replace ineffective links in the formal network. They may also serve as key input points for information distribution. Management may rely upon them for rapid, efficient dispersal of important messages.

Individual engineers may also rely on this for information not otherwise readily available through formal channels, or to distribute information directly to other divisions. This strategy has its risks, however, since those who are the formally designated supervisors and linkages may feel undercut, and therefore resent such abuse of formal procedures. Legitimately, informal communication may lead to the manager being uninformed, and therefore unable to effectively coordinate subordinates' activities. Use of informal channels may prove efficient on one hand, but professionally risky on the other.

Tompkins describes an example of formal networks in technological organizations; NASA—the National Aeronautics and Space Administration.[10] This agency, through the 1960s and 1970s, became one of the most complex organizations on earth. Yet it performed its missions during this time with relative efficiency, even as it constantly grew and changed structure. One of the ways it was able to do so was through attention to, and control of, communication.

Wernher von Braun, the director of the Marshall Space Flight Center (a complex research and development center of which NASA is the parent organization) during this period instituted a system which allowed him to keep informed of the activities of his many divisions. Each Monday, nearly two dozen of his managers sent a one-page memo to von Braun describing that week's activity in their sections. Upon receipt, von Braun read each memo and wrote comments, suggestions or questions. All of the memos were then gathered together, duplicated, and distributed as a package to the managers. Not only could von Braun keep abreast of his organization through this form of upward communication, but managers also received horizontal communication about the activities in other units. On the basis of this sharing of information managers were often able to identify mutual problem areas to provide assistance to other divisions.

Tompkins suggests many factors contribute to the success of the system, but three are perhaps noteworthy here. First, the form of the reporting and the responses was *personal*. When the parent organization NASA tried to implement a weekly memo system, detailed specified forms were developed that eliminated informality. The larger NASA system failed, perhaps because it reinforced a depersonalized perception within that organization.

A second factor was the personal feedback von Braun provided. His notes allowed managers to perceive, almost immediately, their director's response to their activities. They rapidly could adjust their activities accordingly.

Finally, this system worked because the director *required* people to respond to what they read in other sections' reports if there was any way in which one unit manager could assist another. A manager was assumed to have "automatic responsibility for any problem he perceived which fell within his area of competence, regardless of whether or not his laboratory had been given a formal task assignment related to it."[11]

Managers, in turn, began requiring weekly memos from their subordinates in each division. Thus, this relatively simple communication mechanism provided a pervasive network linking units together.

While many of you may not be in a position to institute such a system, it may well be a useful process to suggest or implement if your authority eventually permits. This is but one way in which communication may be managed within the technical organization. It shows particularly well, however, how a little planning and creativity can help structure the organization's communication effectively. Clearly, we think, it also points out how the communication responsibilities of engineers and technologists ought to be delineated just as any other task functions.

## COMMUNICATION LOAD

In Chapter 1 we wrote about how communication problems in organizations are often a result of quality of communication, rather than quantity. In fact, many persons suffer from what is termed *communication overload*: too much information coming in too fast for effective processing. Engineers, particularly as they assume managerial responsibilities, are often inundated with communication demands. When these demands reach high levels, our efficiency in responding to new information drops. To cope with the incoming flow we may select several strategies.[12] These include simply ignoring the backlog of unprocessed inputs (memos, letters, reports), allowing lines to develop (first in, first out), or setting priorities and processing high-priority messages first. These are the typical strategies an individual can adopt. Each has its merits, depending on the situation. In most cases we favor setting priorities. None of these, however, eliminate the problem. Solutions to overload usually require the action of the organization.

The organization's solutions are to provide new permanent staff to handle communication, or to create a "reserve" pool of employees who can fill in as needed in various divisions. Both of these require a commitment of additional personnel, which may not be cost effective.

A final solution, though a difficult one, involves analyzing the necessity of the communication creating the overload. To do so requires reviewing the communication inputs and outputs of each individual in the organization hierarchy. Often, it may be determined that the formal communication system has unnecessary requirements. This may happen because the communication structure has not been planned, *as a system*. Many reporting and message requirements develop as responses to unique crisis situations, or as a matter of convenience to one member (at considerable expense to other individuals). Many reporting requirements may often be eliminated or combined when they are explicitly identified, compared and analyzed.

Organizations and individuals sometimes resort to *shot-gun* techniques of communicating. That is, they send all of their information to everyone to assure distribution to all who "need to know." Such a system is grossly inefficient and wasteful. By overloading everyone with unnecessary inputs which must be acknowledged or at least scanned for relevant information, we reduce the time available for (and the impact of) vital information. For instance, we ignore large parts of our daily mail because so much is useless. No doubt important information slips by us.

The solution to this is, again, somewhat difficult. We believe it is worthwhile, however. The basic requirements of the solution are two-fold. First, the organization members must be encouraged to address their communication selectively only to those with reasonable need to know, rather than to use a shotgun approach. This requires that communicators make a legitimate effort to identify those who need to know specific types of information.

Second, each division or individual should be encouraged to identify the information they require. The first and second steps described here should be compared and adjusted. There is always some risk that absolute information will be inadequately distributed, since it is nearly impossible to accurately anticipate everyone's information needs. Distributing a list of available information may help avoid this; on a weekly (or other realistic interval) basis send either abstracts or titles of your output to all divisions so they may see whether information they need has bypassed them. Modern word processing systems facilitate this type of recording of message output and distribution.

The essence of this type of system is to allow individuals to obtain needed information without having to deal with all the information in the environment. A system like this may be costly, but we think that in many large organizations, which are characterized by high levels of interdependence, this cost may be justified.

This system allows the individuals in the organization to control to some degree, their own communication load. While some individuals may not follow up and request this information (thus reducing efficiency and productivity), we feel the alternative is worse. When individuals receive communication beyond their processing capacity the additional information is usually lost, and therefore useless.

# SUMMARY

It is not enough to merely develop personal skills as a communicator. You must develop the ability to control the communication system you are a part of: to use the system effectively and to adapt it, when you can, to improve your opportunities to use your personal skills.

Communication is a complex process operating within complex systems. We stated at the outset of this book that we could not offer absolute criteria for communication effectiveness. We have offered, we hope, suggestions that have made you examine your own communication, and the communication system around you. From now on we hope that you will be able to make more informed decisions about your communication, and thus improve your effectiveness as a communicator and as an engineer.

Communication is creativity on the part of all who interact. We cannot afford to assume that communication has occurred properly without checking up on the results of our efforts. We cannot afford to leave all of our communication to chance, either as individuals or as organizations. Noise must be reduced. Communicators must develop a wide repertoire of skills, from which they may choose the appropriate words, content, and treatment to match the attitudes and experiences of those with whom they interact. We hope that you will continue to develop new skills and new styles, recognizing that no one strategy of communication will work for "all the people, all the time." We cannot remove from your shoulders the responsibility you have, as much to yourself as anyone, as a communicator. To ignore your responsibility reduces your own and your organization's chances for survival and advancement.

### Questions and Activities

1. Develop a model of the communication system of your organization. Who and what are the critical elements affecting that system?
2. Develop a list of times you and others you know have represented your organization to the public. Were you prepared for these opportunities? Did you know company policy? Did you report back your perceptions of audience response to others in the organization who might find the input useful and enlightening? What could be done by employees such as yourself to enhance the image of your organization and to make your organization responsive to the public's concerns?
3. Which members of your organization seem to maintain the greatest contact with the outside technological environment? How do they maintain these links? Do you regularly use this person as a source of needed information? Do you maintain an effective open relationship with the gatekeepers? If you are a gatekeeper, do you share information freely with colleagues, or use your knowledge as a bargaining chip to control your environment?

4. Make a list of the important information you must receive to be satisfied and effective in your position. What information do you need, but don't receive? Looking at the information you have received in the last several days, which pieces were either unneeded or only peripherally relevant? How could communication be structured to obtain the information you need and reduce unneeded material, while still having it available if needed later?
5. Do you make any attempts to systematically determine the attitudes and gripes of subordinates? How could you structure communication with subordinates to encourage sharing of problems before they become disasters?

## References

1. Goldhaber, G. M. *Organizational communication*, [2nd ed.] Dubuque, Iowa: Wm. C. Brown, 1979.
2. Zelko, H. P. & Dance, F. E. X. *Business and professional speech communication*, [2nd ed]. New York: Holt, Rinehart and Winston, 1978.
3. Rosenbloom, R. S. & Wolek, F. W. *Technology and information transfer.* Boston: Division of Research, Graduate School of Business Administration, Harvard University, 1970.
4. Allen, T. J. "Roles in technical communication networks." *Communication among scientists and engineers.* Carnot E. Nelson and Donald K. Pollock (Eds.), Lexington, Mass.: Heath Lexington Books, 1970.
5. Sherwin, C. W. "Project hindsight: A defense department study of the utility of research." *Science,* 1967, *156*, 1571-1577.
6. See Allen, T. J. "Performance of information channels in the transfer of technology." *Industrial Management Review,* 1966, *8*, 87-98; Allen, T. J. "Communication in the research and development laboratories." *Technological Review,* 1967, *70*, 31-31; and Allen, T. J. "Information flow in research and development laboratories." *Administration Science Quarterly,* 1969, *14*, 12-19.
7. Marquis, D. Q. & Allen, T. J. "Communication patterns in applied technology." *American Psychologist,* 1966, *21*, 1052-1060.
8. Brillouin, L. *Science and information theory.* New York: Academic Press, 1962.
9. Farace, R. V., Monge, P. R., & Russell, H. M. *Communicating and organizing,* Reading, Massachusetts: Addison-Wesley, 1977.
10. Tompkins, P. K. "Management qua communication in rocket research and development." *Communication Monographs,* 1977, *44*, 1-26.
11. *Ibid.* p. 11.
12. Farace, Monge, & Russell. *op. cit.*

# INDEX

Abstracts, writing, 71, 73
Acceptance messages, 95–97
   explicit, 96
   implicit, 96–97
Acronyms, 65
Allen, T. J., 112n, 115
Analogy, 12, 54, 62
   description by, 24–26, 31
   suggestions for using, 26–27
Anxiety, speech delivery and, 54–55
Appendices (in a technical report), 76
Audience
   beliefs, addressing, 32
   identifying one's, 19, 21–24, 48, 58, 60–61
   motivating, 23, 38
   persuading, 31–32
Authorities, citing, 31

Bar graphs, use of, 34–35
Beavin, J., 87, 110n
Beebe, S. A., 101, 105, 110n
Beliefs, addressing audience, 32
Believability. See Credibility
Berlo, D. K., 18n, 43n
   helical model of, 15, 20, 38
Bibliography (in a technical report), 75–76
Brainstorming, 105
Brillouin, L., 116, 122n

Career success and communication, 4–5
Certainty vs. provisionalism, 91
*Chicago Manual of Style, The*, 75–76
Communication(s)
   barriers to, 6–13
   climate, importance of, 98
   complementarity in, 97–99
   concepts, 13–17
   disconfirming, 91–92
   interpersonal, 82–109
   negative, 92–95
   nonverbal, 100
   oral, 44–55, 61–62, 82–109
   overload, 119–120
   with public, 4–5, 113–115
   technical, 19–42, 115–119
   universities with industry, 115
   written, 57–81, 61–62
Communication networks, 32–33, 112, 116–119
Comparison. See Analogy
Complementarity, 97–99
Concepts, explaining, 24–28
Concreteness, importance of, 64
Conflict, 108–109
   managing, 104
   substantive vs. affective, 108
Control vs. problem orientation, 89–90
Conversation
   content, 87
   relationship, 87–88
Credibility, persuasiveness and, 31–32, 53

Dance, F. E. X., 6, 17n, 22, 43n, 49, 56n, 86, 91, 92, 93, 94, 96, 98, 110n, 113, 122n
Decision making, group, 101
Defensive communication, 88–91
Definitions, 28
Delivery (of a speech), 45–46, 54–55
Demographics, 20–21
Descriptions, technical, 28–31, 62
   organizing, 28–29
Detail, importance of, 63
*Directions for Abstractors*, 73
Direct response, 97
Discomforting behavior, 109
Doolittle, R. J., 44

Erickson, H., 3, 17n
Evaluation vs. description, 89, 94
Examples(s)
   description by, 27–28
   importance of, in speeches, 54
Extemporaneous speaking, 47–50
Eye contact, 46, 47, 97. See also Communication

Farace, R. V., 112n, 116, 119
Flexibility, importance of, in oral presentations, 44–45

## Index

Followup, importance of, 99, 100
Footnotes, use of, in a technical report, 75–76
Frame of reference, 48
Future shock, 13, 23

Gelderman, C., 57
Gender, 21
Gibb, J., 89–91, 93, 110n
Goldhaber, G. M., 113, 122n
Government, communication and, 5
Graphics, communication and, 5
Graphics, importance of, 32–38, 62
Graphs, use of, 34, 35, 36, 37, 39
Group(s)
    advantages and disadvantages of, 101
    interaction, 102–104
Guetzkow, H., 108, 110n
*Guide to Source Indexing and Abstracting of Engineering Literature, The*, 73
Gyr, J., 108, 110n

Heading(s)
    in a letter, 68
    in a written report, 41
Hopper, R., 102

Illustration list (in a technical report), 74
Impartiality, persuasion and, 31–32
Imperviousness, 94, 97
Impromptu speaking, 51–54
Information
    absolute, 116
    adapting, 82
    distributed, 116
    sharing, 101–102
Inside address (of a letter), 69
Interpersonal communication, 82–109
Interrupting, 94
Interview(s)
    concluding, 99
    following up, 99
    information-gathering, 83–87
    and structure, 83, 84–85
Introduction (of a technical report), 74
Irrelevant response, 94–95, 97

Jackson, D., 87, 110n
Jargon, 11–12, 64, 66
Johnson, R. J., 108, 110n
Jones, J. S., 43n

Key words, importance of, in reports, 71, 73

Larson, C. E., 91, 92, 93, 94, 96, 110n
Leary, T., 88, 110n
Letters, transmittal, 66–68
    guidelines for writing, 68–71
Listening, active, 99–100

McLuhan, M., 32, 43n
Majority rule, problem solving and, 104
Managers, engineers as, 1–3, 5
Manuscripts, speaking from, 45–46
Marquis, D. Q., 115, 122n
Masterson, J. T., 101, 105, 110n
Memorization (of oral reports), 47
Memos, 76–81
    functions of, 77
    preparing, 78–81
Messages, treatment of, 38–41
Monge, P. R., 116, 122n
Motivation, audience, 23, 38

Nebergall, R. E., 23, 43n
"Noise," 14–16, 20, 21, 24, 121
Nonverbal communication, 100

Objectivity, importance of, in persuasion, 31–32
Oneupmanship, 98, 109
Oral communication, 44–45, 61–62, 82–109. See also Delivery (of a speech)

Page, P., 4, 17n
Perelman, S., 4, 17n
Persuasion, 23
Pie charts, use of, 35, 36
Pirsig, R., 12, 18n
Problem
    definition, 102, 105
    solving, 102–104, 108–109
Process description, 29, 49
    principles for, 30
    visual aids and, 34
*ProCom* series, 44, 57
Proofreading, importance of, 68, 70
Public, communication with, 4–5, 113–115
Purpose, 47–49, 57–58

Questions
    answering impromptu, 52
    asking, in interviews, 84–85
    closed- and open-ended, 85

Reading aloud (one's writing), 59–60
Redundancy, effective use of, 39, 41
Reflective Thinking Format, 104, 105–108
Rejection messages, 92–95
  implicit, 94–95
Relationships, interpersonal, and interviews, 86–88
Reports
  oral, 44–55
  technical, 66–78
Rewriting, importance of, 59
Rosenbloom, R. S., 115, 122n
Russell, H. M., 116, 119, 122n

Salutation (of a letter), 69
Scripts. See Manuscripts
Shannon, C., 14, 18n
Sherif, C. W., 23, 43n
Sherif, M., 23, 43n
Sherwin, C. W., 115, 122n
Sieburg, E., 91, 110n
Spelling, correct, importance of, 68, 81
Stance, communicator's, 32
Strategy vs. spontaneity, 90
Style, written, 61–66
  impersonal, 63–64
  objective, 63–64
*Style Manual*, G. P. O., 76
Summarizing, 41
  interviews and, 85

Superiority vs. equality, 90–91
Symmetricality in communication, 97–99
Systems, communication, 112–113

Table of contents (in a technical report), 74
Tangential response, 95, 97
Technical communication, 19–42, 11–121.
  See also Reports, technical
*Thesaurus of Engineering Terms*, 73
Title page (of a report), 71
Toffler, A., 13, 18n
Tompkins, P. K., 81n, 118, 122n
Tone, 46

Visual aids, 32–38

Watzlawick, P., 87, 110n
Weaver, C., 100, 110n
Weaver, W., 14, 18n
Weisman, H., 28, 43n
Wilmot, W. W., 98, 110n
Wirkus, T., 17n
Wolek, F. W., 115, 122n
Word choice, 65–66
Writing, effective, 58–61. See also Communication, written

Zelko, H. P., 6, 17n, 49, 56n, 113, 122n

4411